BIOMEDICAL SENSORS AND SMART SENSING

A Beginner's Guide

Primers in Biomedical Imaging Devices and Systems

Series Editor: Nilanjan Dey

BIOMEDICAL SENSORS AND SMART SENSING

A Beginner's Guide

AYAN KUMAR PANJA

Computer Science, Institute of Engineering and Management, Salt Lake, Kolkata, India

AMARTYA MUKHERJEE

Institute of Engineering and Management, Salt Lake, Kolkata, India

NILANJAN DEY

Department of Information Technology, Techno India College of Technology, Rajarhat, Kolkata, India

ELSEVIER

ACADEMIC PRESS

An imprint of Elsevier

Academic Press is an imprint of Elsevier
125 London Wall, London EC2Y 5AS, United Kingdom
525 B Street, Suite 1650, San Diego, CA 92101, United States
50 Hampshire Street, 5th Floor, Cambridge, MA 02139, United States
The Boulevard, Langford Lane, Kidlington, Oxford OX5 1GB, United Kingdom

ISBN 978-0-12-822856-2

For information on all Academic Press publications
visit our website at https://www.elsevier.com/books-and-journals

Publisher: Mara Conner
Acquisitions Editor: Carrie Bolger
Editorial Project Manager: Andrea R. Dulberger
Production Project Manager: Prasanna Kalyanaraman
Cover Designer: Mark Rogers

Typeset by STRAIVE, India

Working together
to grow libraries in
developing countries

www.elsevier.com • www.bookaid.org

Contents

Author's biographies

Ayan Kumar Panja is assistant professor of Computer Science at the Institute of Engineering and Management, Salt Lake, Kolkata, India. He holds a Master of Technology from the University of Calcutta. His primary research interest includes localization, sensor network and sensor cloud architecture, and machine learning. He has written numerous papers and a book in the field of data gathering and load balancing in sensor cloud. He is the organizing committee member of several international conferences and has played an active role as the co-convenor of the conferences.

Amartya Mukherjee is assistant professor at the Institute of Engineering and Management, Salt Lake, Kolkata, India. He holds a bachelor's degree in computer science and engineering from West Bengal University of Technology and a master's degree in computer science and engineering from the National Institute of Technology, Durgapur, West Bengal, India. His primary research interest is in embedded application development, including mobile ad hoc networking, delay-tolerant networks, and Internet of Things and machine learning. He has written several research articles, books, and book chapters in the field of wireless networking and embedded systems in various journals and publication houses, such as Springer, CRC Press (Taylor & Francis Group), Elsevier, IEEE, World Scientific, and IGI Global. His book titled *Embedded Systems and Robotics with Open Source Tools* is one of the bestselling books in the field of embedded application development. (Website: https://sites.google.com/iemcal.com/amartyamukherjee/home.)

Nilanjan Dey is an associate professor in the Department of Computer Science and Engineering, JIS University, Kolkata, India. He is a visiting fellow of the University of Reading, United Kingdom. He is an Adjunct Professor of Ton Duc Thang University, Ho Chi Minh City, Vietnam. Previously, he held an honorary position of Visiting Scientist at Global Biomedical Technologies Inc., CA, United States (2012–15). He was awarded his Ph.D. from Jadavpur University in 2015. He has authored/edited more than 90 books with Elsevier, Wiley, CRC Press, and Springer, and published more than 300 papers. He is the Editor-in-Chief of the *International Journal of Ambient Computing and Intelligence* (IGI Global), Associated Editor of *IEEE Access*, and *International Journal of Information Technology* (Springer). He is the Series Co-Editor of Springer Tracts in Nature-Inspired Computing (Springer), Series Co-Editor of Advances in Ubiquitous Sensing Applications for Healthcare (Elsevier), and Series Editor of Computational Intelligence in Engineering Problem Solving and Intelligent Signal Processing and Data Analysis (CRC). His main research interests include medical imaging, machine learning, computer-aided diagnosis, data mining, etc. He is the Indian Ambassador of the International Federation for Information Processing—Young ICT Group and Senior member of IEEE. (Website: https://sites.google.com/view/drnilanjandey/home.)

Preface

Sensing in the healthcare system is the foremost essential and emerging aspect. In the case of condition monitoring of the human organs, tissues, and muscles, the sensors have a wide range of usage. Second, computer-aided diagnosis plays a vital role in detecting and predicting complex diseases with great accuracy. The introduction of the high-end sensing devices and micro-electro mechanical systems (MEMS) has made a step forward to build and deploy precise health monitoring systems and ubiquitous computing platforms for medical data processing. There are several messaging protocols that are also involved with it, which performs critical message transfer through the intelligent network system.

The fundamental objective of this book is to provide the reader an insight into the current advancement of biomedical systems and their fundamental features such as sensing, signal processing, data acquisition, data agglomeration, data mining, and message transfer within biomedical networks. Chapter 1 mainly provides an overview of sensors and their characteristics and also provides an insight into biopotential and signals. Chapter 2 covers the sensing and data gathering techniques used in the biomedical system design. Modern applications such as a smart shoe, cloud-based ECG monitoring, and EEG-based sentiment analysis have been presented in the chapter. In Chapter 3, signal processing and imaging have been emphasized followed by the body sensor network. In Chapter 4, the authors have briefed and explained the concepts of machine learning and feature modeling with respect to biomedical signals. Various supervised and unsupervised learning methodologies have also been discussed in the chapter. Chapter 5 emphasizes the medical cyber-physical systems, which cover the ubiquitous sensing approach along with the low-latency communication methodology. Chapter 6 describes the overview of smart perishable food and medicine management. Chapter 7 discusses data gathering and cloud computing methodology for healthcare. This book itself is written in such a way that the readers who are not much familiar with the concept of sensing for health-care application and willing to perform research in this domain can easily understand the fundamental concepts. The book will definitely serve as a guide to the graduate research students who can gain a certain level of understanding about intelligent wireless sensor networks.

Ayan Kumar Panja

Amartya Mukherjee

Nilanjan Dey

Acknowledgments

We thank the researchers for their enormous research efforts toward the modern sensor network and intelligent network applications. We also thank our students and co-researchers. Last but not least, we thank our wives, children, and family members for their continuous support.

Ayan Kumar Panja

Amartya Mukherjee

Nilanjan Dey

About the book

Modern biomedical systems use various computational techniques for smart and intelligent decision making. The purpose of such systems is to design and develop a seamless infrastructure that produces a fast and accurate result in a mission-critical healthcare scenario. The system must comprise various communication and computational frameworks that ensure a high degree service in terms of processing and data management. The chapters of this book have been arranged in such a way that the fundamental concept of biomedical sensing is covered followed by various sensing aspects along with the data aggregation and preprocessing techniques. IoT-based smart home management and patient management is another key aspect that has been covered in the book.

Chapter 1 provides a fundamental overview of biomedical sensing and biomedical systems. The physical characteristics of the communication systems are a main challenging aspect of the chapter. The chapter also covers the fundamental aspect of signal processing and its challenges in the context of medical systems. Various sensing aspects such as magnetic sensing, electrical sensing, and acoustic sensing and their features have been illustrated as well. Numerous signal entities and parameters that affect the signal parameters are mainly emphasized in the chapter.

Chapter 2 provides a clear idea about the signal and the noise and artifacts related to the sensor. It also focuses on the calibration and the error elimination scenario. The chapter highlights the measurement techniques. The fundamental feature of the electrocardiogram system and smart ECG application is also discussed in the chapter. Also, a major concept of ultrasound-based blood flow monitoring has been incorporated in this case. The implementation of EEG and ultrasound-based fetal growth observation technique and its basics has also been discussed. Obstetrical sonography is another key concept that has been discussed in the chapter.

Chapter 3 mainly considers the signal processing aspects. Various intelligent techniques are adopted and evolved to create feasible and optimized algorithms. The key concept of time-series analysis using the ARIMA model has been discussed. Another important concept that has been implemented is biomedical image analysis. In this scenario, various image processing techniques have been adopted. The wearable and implantable sensors and the data classification techniques have been implemented as well.

Chapter 4, on the other hand, describes the fundamentals of the machine learning technique that has been adopted for data analysis and health informatics systems. Various classification techniques in ECG signals and images have been discussed as well. Wearable health care and WBAN are the key features that are emphasized in the chapter.

Chapter 5 presents the medical cyber–physical systems. The use of various networking technologies such as 5G, Wi-Fi, and IoT is covered in the chapter. The practical scenario and the implementation of vehicular technologies such as UAV-based healthcare have been discussed. The real-life application of several communication technologies in the prevailing COVID-19 pandemic situation is also a key topic of interest of the chapter.

Chapter 6 mainly provides an overview of the perishable food and medicine management. Food and medical supply chain management and their networking methodologies are the key aspects of the chapter.

Finally, Chapter 7 primarily discusses the cloud computing paradigm in the healthcare system. The use of cloud computing methodology is not new in the healthcare and biomedical system. The chunk of data that is gathered using sensors is often redirected to the cloud infrastructure using IoT networks and protocols. An illustration of the cloud virtual environment and the system orchestration has been implemented and discussed in this segment.

Nowadays the emerging technologies produce crucial sensing and health management ecosystems that are perhaps the core building block of future healthcare and informatics management.

CHAPTER 1

Introduction

Contents

1.1 Biomedical sensors and system overview

The sensors form the core part of the biomedical systems [1]. A simple example of a biomedical system will be a digital thermometer, measuring the temperature of an individual and processing it digitally. The duty of any biomedical sensor is not just sensing but also converting the biological raw data into digital signals. Hence unlike normal sensors, the biomedical sensor itself is the interface between a living entity and the digital processing system. In general, if we classify sensors, we have physical sensors, chemical sensors, biosensors, etc. The biomedical sensor is the agglomeration of all the above-mentioned types providing a plethora of sensing applications.

To have quantified measurements physical sensors like piezo-electric sensors, temperature sensors, photoelectric sensors, acoustic sensors, etc. are widely used. Chemical sensors measure values about objects that are more of chemical nature like humidity sensors, various electrodes, optical gas sensors, etc. Biosensors, on the other hand, are a combination of both physical and chemical sensing together, some of the examples are gravimetric sensors, pyroelectric sensors, optical photoelectric sensors. Biomedical measurement [2] is a guiding technology [3] in the collection and processing of information about the medical domain and is directly related to the research of biomedical sensing technology, biomedical measurement methods, and electrical-electronics measuring systems. Therefore the research carried out in biomedical measurement has a direct effect on the design and application of sensors and medical instruments. An overview of a biomedical system can be a simple pressure sensor for blood pressure measurement, electrocardiogram (ECG) [4], electromyography (EMG), photoplethysmography (PPG), etc. There are various systems that can also detect molecules, enzymes, and measure many such biological fluids paving the way to a complicated diagnosis of various diseases (Fig. 1.1).

1.2 Physical characteristics

The core part of any system is the sensors. The development of a biomedical system application involves many types of sensors. The behavior of any sensor can be classified as stable or static and dynamic. Static characteristics can be estimated after all effects that last for a small time that have stabilized to their final or steady state, while dynamic characteristics describe the sensor's transient properties, and with respect to time-varying input, they can be measured. In simpler terms, if we consider a measurement in a stable or an environment that changes less frequently the measurements are categorized as static. For biomedical sensing few stable characteristics of sensors are as follows:

(1) *Lifetime*, which is the measure of time the sensors remain sensitive in a stable condition of temperature, pressure, and humidity maintained at some fixed ideal values.

(2) *Sensor sensitivity* is the measure of the ratio between the output signal change to the measure of input property.

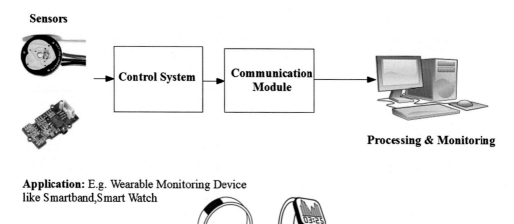

Fig 1.1 Biomedical system application overview.

(3) *Range,* the range of sensing that is the maximum detectable range of the sensor.

(4) *Selectivity,* a measure of a component by a sensor in the presence of another.

(5) *Error and accuracy* is defined as the difference between the measured value and the actual value. Accuracy defines how correctly the sensor output compares to the actual value.

When we consider the time-varying signals the response characteristics are called dynamic characteristics. The response comprises two parts, short-time signals and finite sinusoids. Short-time signals are bursts of events that are generated for a short time. And the steady-state response is the output state which is expressed as finite states of sinusoids. When we are considering biomedical sensors, the majority of sensor data are functions of time; in order to obtain the information from the physical body, the biomedical sensors should have a combination of both static and dynamic sensor characteristics. Now the signals produced can be categorized and accordingly the devices can be manufactured to gather the relevant data. We can broadly categorize them as electrical, magnetic, and mechanical signals.

1.2.1 Magnetic sensing

The magnetic signals from the magnetic field produced by various tissues, fluids, etc. in our body can be recorded for data gathering. The term used for such types of signals is called "*Biomagnetic signals.*" The important biomagnetic measurement from the human body is Magneto-cardiogram (**MCG**), Magneto-encephalography (**MEG**), and Magneto-myogram (**MMG**). Magnetic sensors work on the principle of the change of magnetic moment that takes place in a magnetic material, when submitted to a magnetic field a change in temperature or energy occurs. Hence the variable is utilized to study the changes that are occurring. The materials that are selected for creating the sensors have to be done in a manner such that there is minimal interaction between the magnetic material and the organic compound. The organic matter may be harmed if the most popularly used materials, which are magnetic (transition metals and their alloys), are not biocompatible, although iron is biocompatible. The organic fluids have ions and several organic radicals which are corrosive for the magnetic materials that are used for sensing. One of the solutions to the problem consists of covering the magnetic material with a thin layer of another biocompatible material like gold, titanium oxide, silica or alumina, etc. This is about the metallic part of the sensor. For the nonmetallic parts of the sensor, it has to be made with biocompatible polymers or ceramics such as Teflon, etc.

1.2.2 Electrical sensing

Electrical signals are present throughout our bodies. The nervous and circulatory systems are one of the major sources of signal carriers of electrical impulses. The most popular procedures of electrical signals in our body are Electrocardiogram (**ECG**),

Electroencephalogram (**EEG**), and Electromyogram (**EMG**). The ECG is the simplest of the detection of electrical signals from our heart to check its condition. To measure the electrical activities in our brain we utilized the EEG procedure. EMG, on the other hand, is mostly utilized to gather electrical signals from our body muscles. The sensors used for the above-mentioned approaches are electrodes. Some of the popular electrodes utilized for biomedical instrumentation are Microelectrodes, Body surface electrodes, Subdermal Needle electrodes, etc.

1.2.3 Acoustic sensing

The simplest of the acoustic sensors is a stethoscope used to hear the heartbeat. The acoustic stethoscope can be utilized to record the heartbeat and the waveform can be recorded. Acoustic sensors that record sound can also be utilized to record various respiratory sounds which can be recorded for diagnosis of various diseases. Some of the popular sensors utilized are piezo-element, acoustic accelerometers, or a simple microphone. The materials used for creating the sensors are lithium niobate and lithium tantalite-based materials. Lithium tantalite is highly used for the production of acoustic sensors which are used in biomedical applications as they possess unique optical, piezoelectric, and pyroelectric properties. The biosensors used to detect the physical and chemical information inside an individual's body employ acoustic or mechanical waves for the detection.

1.3 System and signal

The objective of signal processing involves the measurement of various signals pertaining to a system, in the case of biomedical signal processing it involves the measurement of signals pertaining to the human system. Now if we are to define a system, a system is a group of homogeneous and heterogeneous units that work toward a common purpose. The human body is composed of many subsystems that enable us to do myriad things. The coordinated action of the subsystems works toward a single objective of the whole system. The subsystems involved in our human body are the circulatory system, nervous system, skeleton system, respiratory system, excretory system, digestive system, muscular system, and reproductive system. The physiological activities ranging from protein sequences to nerve conduction, cardiac rhythm, tissues, and organ images, etc. all are sources of information that are measured for various diagnoses of the health of an individual.

Biomedical measurements have diversified applications. In the process of biomedical measurement, it is very crucial that the measurement procedures employed should have less interference and low noise. Furthermore, the research carried out should have analog circuits as well as digital, processing units, measurement systems, etc.

1.3.1 Measurement

The sensors gather signals which have to be preprocessed and have to be converted into a usable quantifiable form for its usage. A simple measurement system can be modeled into a form containing sensors at the lower layer followed by a signal modulation unit, filter circuits, and signal convertors (A/D and D/A). A simple signal processing circuit is given in Fig 1.2 explaining the stages involved in the process. The sensors are the primary devices used for data acquisition; the gathered data from the environment are passed through the various circuits. The conversion circuit is responsible for converting the gathered raw data into electrical impulses. An amplifier and filter circuit is very much required to amplify the signal and filter out the noise from the signal. The analog signal is passed through an A/D converter to transform it into a digital signal. The digital signal is a series of numbers, discretized in both the amplitude and time domain. The steps involved in the A/D converter are shown in Fig 1.3.

The most widely used analog-to-digital conversion method is pulse code modulation (PCM). In the PCM procedure, the analog signal is passed through the following three steps in order to convert to digital data.

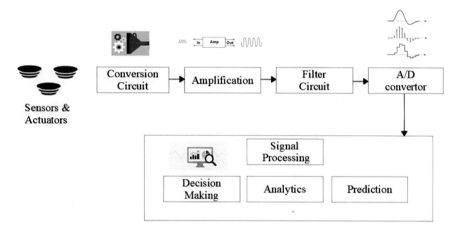

Fig 1.2 Signal processing and analysis.

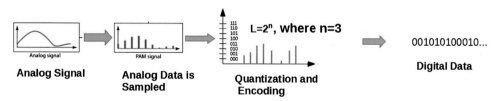

Fig 1.3 Pulse code modulation.

(1) Sampling: Converts analog signal into the discrete signal.
(2) Quantization: Converts discrete signals into digital signals.
(3) Encoding: Encodes the digital signal with binary values.

The continuous signal is converted into discrete time series through the sampling process. The quantization procedure allocates the amplitude values of each sample within the set of discrete determined values. The filter circuit is connected before the A/D convertor with the objective to eliminate the high-frequency components so the sampling errors get decreased.

If we are to classify the data measurement and gathering procedure, there are three ways in which measurements are performed:
(1) In vitro: Measurement performed within the body.
(2) In vivo: Measurement performed outside the living organism.
(3) In situ: Measurement performed directly at the point of interest. Something which is a combination of both In vitro and In vivo.

Similarly, another form of classification of measurement procedure can be:
(1) *Invasive*: Pervading inside the body by causing destruction to cells.
(2) *Noninvasive*: Nondestructive procedures can include measurement inside the body as well as outside.

Biocompatibility is a very serious issue for the sensors that are used in long-term implantable mode. It is very important that the sensors implanted should be nonirritating, nontoxic, and noncarcinogenic in nature. The chemical composition of the organism or the individual's tissues has to be studied properly before the design of the sensors. Some of the examples of biocompatibility are as follows:
(1) Histocompatibility: Our body's immune system attacks any foreign materials that enter the body. Thus the sensors should be histocompatible in nature.
(2) Blood compatibility: The interaction of the sensors with the blood has to be studied and analyzed. The sensors used in the cardiovascular system are in direct contact with the blood in an individual's body. Many different reactions can be triggered by the interaction of the two. Hence the material used for the sensor design should be blood compatible.

1.3.2 Biopotentials

The building blocks of our body are the cells. The cells bind together to form the tissues and the tissues together form the organs. The whole human system is formed by the coordinated effort of the organ subsystem. Nearly 100 trillion cells are there in our body, and 50 million of them get destroyed and created every day. The important thing that keeps the cells active and alive to perform all of the tasks is called the metabolism procedure. The metabolism can be classified into two terms—anabolism and catabolism. Catabolism refers to energy gathering by breaking down molecules, while anabolism refers to the synthesis of all compounds needed by the cells. The metabolic pathways depend upon nutrients that are broken down in order to produce energy. The energy is utilized to synthesize new proteins, nucleic acids, etc. The energy is exchanged across the cell wall of the cell. The mitochondria as we call it are the powerhouse of the cell and are responsible for the oxidation process and the production of Adenosine Triphosphate (ATP). The cell membrane separates the intracellular fluid, i.e., the cytoplasm from the interstitial fluid or the tissue fluid. The cell membrane is a permeable membrane that allows the transmission of ions from outside to inside and vice versa. A block diagram of the cell membrane, cytoplasm, and interstitial fluid is depicted in Fig 1.4. The metabolism procedure is the transport mechanism of glucose and oxygen which gets oxidized in the mitochondria and end products are CO_2 which is released. The cytoplasm and interstitial fluids contain ions such as potassium, chlorine, sodium, etc. Cell membrane as it is semipermeable allows the movement of potassium ions and sodium ions to flow from higher concentration to lower concentration. If the outside concentration or potential is high, then K+ ions come inside the cytoplasm through diffusion. As a result, we lose positive ions on the outside; hence it goes into a state of the negative potential of -60 to -90 mV across the cell membrane. An equilibrium is reached, that is repulsing the K+ ions to go out and a ratio (K+: Na+) of 10:1 is maintained (Fig 1.5).

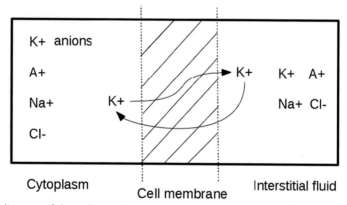

Fig 1.4 Block diagram of the cell membrane, cytoplasm, and interstitial fluid.

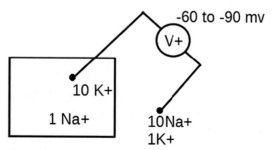

Fig 1.5 Block diagram of equilibrium potential of cell.

Now to perform the measurement, we need the ions to flow and this only happens when a stimulus or an external potential is applied across the cell membrane. This electrical impulse causes the permeability of the cell membrane to change and the ions to flow (Na+ flows inside and K+ flows outside). This procedure is called depolarization and the procedure continues till the potential of the cell reaches 20 mV from −90mV. After reaching the peak potential a repolarization starts which is the reverse process and the K+ ions are absorbed inside the cells through the cell membrane till an equilibrium potential of −90 mV is reached. For muscle cells, the depolarization phase has an extra phase associated with it called the contraction phase. These waveforms or electrical impulses as it travels through the body are utilized to perform all the bodily activities. The signal that they never carry is these electrical impulses. The same impulse also drives the heart and all the other organs. The electric current flows from the cytoplasm through the cell membrane to the interstitial fluid. These potentials are called "biopotential" (Fig. 1.6)

Some of the characteristics of the "biopotential" are as follows:

(a) The electrochemical activities that happen through the cell membrane give rise to the current which in a coordinated manner drives the whole of the body and is called "biopotentials."

(b) The cells are ionic batteries that happen with the movement of the ions or charged particles. Although the procedure is slower than electronic circuits.

(c) The biopotential gives rise to waveforms and this, in turn, is recorded from a large number of cells. These waveforms are ECG, EEG, etc. which are finally used for data gathering and processing in biomedical application and discussed in the latter part of the chapter.

The collection of waveforms can be done in many forms and ECG is one of the popular waveforms recording the electrical signals from the heart through the electrode. A demonstration of ECG data collection with AD8232 ECG monitor sensor and Raspberry Pi connection is shown in Fig. 1.7.

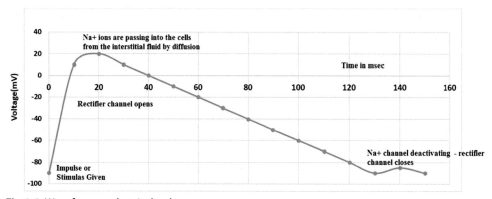

Fig 1.6 Waveform or electrical pulse.

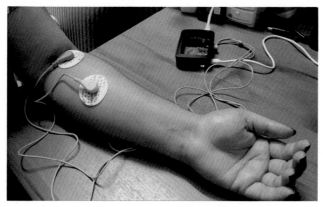

Fig 1.7 ECG monitor connected with Raspberry Pi to collect ECG raw data.

1.4 Sensor characteristics

The performance of the sensor is measured by the characteristics of the sensors. Broadly, we classify it as static and dynamic characteristics. The static characteristics are measured under a stable state, that is when no external force is exerted on the sensors. In the stable state, the measured characteristics include zero drift, lifetime, sensitivity, and sensor range.

Drift or zero drift is the deviation from the actual reading of the sensor when the sensor is in a stable state. In the majority of the cases if there is no input signal then the output should be at zero reading.

Lifetime, sensor lifetime is the duration of the time during which the sensors remain active under ideal stable condition.

Sensitivity refers to the unit change or the gradient of the system in stable conditions. That is with respect to the unit change in input how much change in the output is taking place.

Sensor range, the range is estimated by the accuracy required determined by the lower and upper limit of input and output.

Dynamic characteristics refer to the response of the sensors with respect to time and input. The dynamic characteristics during the period of the sensor are used from its initial state of usage to its final state. As the sensor starts its usage, the sensing system goes through various phases like rising time, peak time, and settling time. The rising time gives the steady-state response time going from 5% to 80% sensitivity. The peak time refers to the time required for the sensor to reach peak accuracy. The settling time is the time when the sensor's response value settles down to a steady-state value. Fig 1.8 gives us an overview of the three main dynamic characteristics of the sensors.

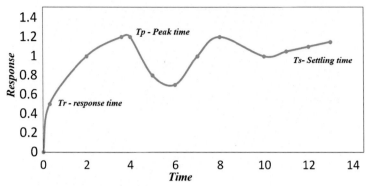

Fig 1.8 Dynamic characteristics of sensors.

1.4.1 Sensitivity of the sensor

Sensors' sensitivity is the gradient change or the ratio of the change in the output with respect to the input parameter. If the sensors have high sensitivity then a menial change in the input might cause a drastic change in the output. A simple overview of an example change will be as follows. Say a temperature sensor has a sensitivity of 15 mV/C; this implies that the output change of 15 mV will occur for every 1°C change in input temperature. Now if the calibration line is linear, then we have a constant sensitivity. Again if the sensitivity varies with the input then the calibration is nonlinear. Hence it is very important for proper designing and calibration of the biomedical sensors for their proper functioning.

1.4.2 Linearity

In an ideal case, the sensor is said to give a certain output value; in normal situations the deviation of output characterization from the ideal case is measured with a measure curve expression as depicted in Fig. 1.9. Linearity is represented as a percent of the actual full-scale output, i.e. the major deviation from the calibration point of the sensor from the corresponding point on the ideal case scenario. Let "*l*" denote the linearity parameter. D_{max} is the maximum divergence from the actual value, it can be positive as well as negative and Y full-scale output.

$$l = \frac{D_{max}}{Y_{FS}} \times 100\%$$

1.4.3 Sensing errors

Errors are expressed as the difference between the actual value and the sensed value. The reason for the occurrence of the errors is due to various environmental conditions and inherent drifts caused by the sensor components.

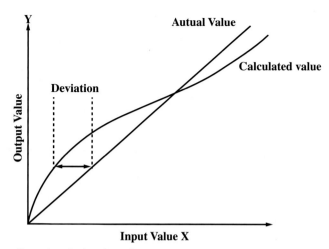

Fig. 1.9 Actual reading v/s calculated reading.

$$\text{Absolute error} = \text{collected measurement} - \text{actual value}$$

$$\text{Relative error} = \frac{\text{absolute error}}{\text{actual error}}$$

The majority of the errors in any processing system are a combination of three basic parts. In order to improve the sensing performance of the system, the precision of the measurement parameter has to be increased. In the biomedical application, the data gathered from the sensors are the first thing that gets propagated throughout the entire system. Hence the accuracy of the sensor is one of the major important things that have to be increased. The design of the sensors includes various heterogeneous components dealing with energies from chemical, biological to mechanical, and electrical. Hence it is difficult to provide a system that can work at the same pace with all the subcomponent parts. In some of the actual applications, there are some aspects that do not have influences on a particular signal. Hence it is not always necessary to focus completely on the design perspective of the sensors. Various filtering procedures like simple averaging and other techniques are used to either mask the effect of the errors or to filter out errors from the system.

1.5 Biopotential signal monitoring and biosensors

In the present era it is important to build sensors that are smaller in size and the present trend also follows the miniaturization of sensors. Biopotential as discussed before are the current flows and measuring them is essential to monitor various body conditions. To measure biopotential some form of electrodes are essential. And with the advancement of technology the integration of such electrodes and devices is being done with the objects that we use daily, for example garments, watches, glasses, etc. One of the evolving technologies is the development of smart textiles. Combining the electrical, thermal, and mechanical properties of the sensors with the garments allows the monitoring and addition of different functionalities. Fabrics merged with sensors by the process of electrode plating, deep coating, physical vapor deposition, etc. are some of the popular processes.

Biosensors [5,6] are devices used for analyzing and devising a tiny biomolecule that provides the understanding for a particular analyte. It is a challenging approach to know the mechanism of biochemistry and the reactions through direct invasive measurements. It is also very difficult to analyze the signals collected from the complex environment inside the organism. An analyte is a chemical substance whose constituents are measured and analyzed. Biosensors contain transducers to collect and quantify the binding taking place between the analyte and the molecule used for detection. Biosensors have a wide range of usage from various detection of bacteria and viruses, sensing the environment, etc. (Fig. 1.10)

Biosensors consist of three components (Fig. 1.11): receptors, a transducer, and a computing or processing system. The baroreceptors are designed with the proper specification to recognize the target analyte. A chemical and biological procedure is utilized for the recognition of the analyte. A biosensitive part or layer is attached to the sampling component. Popular bioreceptors consist of nucleic acid reaction, enzyme interaction, antibody bioreceptors, etc.

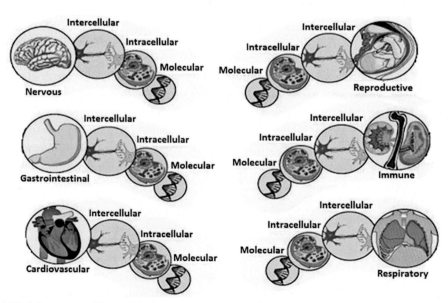

Fig 1.10 Subsystems of human body.

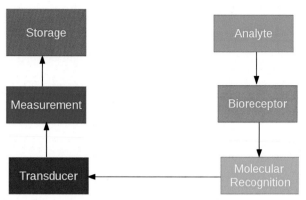

Fig 1.11 Components of bioreceptors.

Antibody bioreceptors or popularly called immune sensors are mainly used for the detection of antigens. They are receptive to the antigen and antibody interaction. Antibodies are heavy molecular structures and are mainly produced by the response of the immune system toward the foreign particles called the antigens. The antibody binds with the antigen and is able to detect and predict it. Now the bioreceptive antibodies are used for detecting the foreign particles or the analyte and the gathered information is finally processed. Enzyme bioreceptors are widely used as enzymes that act mainly as catalysts without being absorbed in the reaction process. Enzymes are natural proteins and are majorly used bioreceptors for biocatalytic recognition. For selective determination and analysis of various types of analytes, nucleic acid-based bioreceptors are also popularly used. Two single-strand DNA are combined to form a double-strand DNA structure which is used in a nucleic acid biosensor that selects the nucleic acid as biological recognition element. Peptide nucleic acid (PNA) is a popular class of nucleic acid used as a bioreceptor.

1.6 Conclusion

The chapter gives a brief introduction to biomedical signals, signal processing, and systems. The chapter discusses biopotentials and how they are measured from an individual's body, giving insights into EEG, ECG, and EMG signals. The chapter also discusses how biopotential signals are monitored and also giving insights on biosensors.

References

[1] M. Engin, A. Demirel, E.Z. Engin, M. Fedakar, Recent developments and trends in biomedical sensors, Measurement 37 (2) (2005) 173–188.
[2] P. Wang, Q. Liu, Biomedical Sensors and Measurement, Springer Science & Business Media, 2011.

[3] L. Schwiebert, S.K. Gupta, J. Weinmann, Research challenges in wireless networks of biomedical sensors, in: Proceedings of the 7th Annual International Conference on Mobile Computing and Networking, 2001, July, pp. 151–165.

[4] K.J. Gilhooly, P. McGeorge, J. Hunter, J.M. Rawles, I.K. Kirby, C. Green, V. Wynn, Biomedical knowledge in diagnostic thinking: the case of electrocardiogram (ECG) interpretation, Eur. J. Cogn. Psychol. 9 (2) (1997) 199–223.

[5] J.S. Schultz, Sensitivity and Dynamics of Bioreceptor-based Biosensors, 1987.

[6] F. Scheller, F. Schubert, Biosensors, Elsevier, 1991.

CHAPTER 2

Sensing and data gathering methodology

Contents

2.1 Signals and noise of sensors

In the world of medical sensing, numerous sensors take part in sensing and monitoring the various critical parts of the human body. The sensors like ECG, EEG, PPG, and EMG are some of them. While sensors are sensing the physical quantity, it is crucial to take the data flawlessly. When a sensor senses a value, it converts into an electrical quantity. That electrical voltage level is known as a signal. In some cases, the signal produced by the sensors is continuous. Such signals are also known as continuous or analog signals. In other cases, some of the sensor data may be discrete. Data generated by the digital sensors are mostly of such kind. Another major kind of component related to signal is the noise. The signal and the noise are perhaps uniquely defined by their physical nature. For example, when we take ECG of a patient if some external voltage or leakage current appears due to bad ground, the signal may be interfered by those voltage components that result in noise in the signal. This problem also happens due to muscle noise that appears in our body as well as forms artifacts. Such kind of noise perhaps cannot be removed even with narrow-band filtering. In real-world situations, there is no way to distinguish the noise from the actual signal but depending upon the intuition of the observer and the detailed knowledge of the measured entity and the possible disturbance generally helps to distinguish between signals and noise. The major formulation to identify the effectivity of a signal is to find out the signal-to-noise ratio. The SNR value can be determined by the pick-to-pick or RMS (root mean square) of the base signal divided by the noise imposed on it. In a practical scenario, it is considered in a limited frequency range. The unit is considered as dB. The expression of the SNR can be written as:

$$SNR = \log_{10} \frac{S}{N} \tag{2.1}$$

Here, S and N define the power of the base signal and the noise component, respectively.

2.1.1 Various classes of noise

In a biomedical sensing environment, various class of noises can be considered. Some of them are like thermal noise, interference, artifacts, and $1/f$ noise.

The thermal noise is generated due to random thermal agitation. The power density of the same is distributed within an entire frequency range, and the power is proportional to the temperature. If we consider the frequency range as Δf and the noise $\nu(t)$, then we can derive a relation as

$$\overline{\nu(t)^2} = 4 \times \kappa\, T\, R\, \Delta f \tag{2.2}$$

Here, κ is the Boltzmann constant and the value is 1.38064×10^{-23} J/K, T is absolute temperature, and R is the resistance of the medium.

The interference, on the other hand, is caused by a physical, electrical, mechanical, or chemical event in the environment. The natural phenomenon like thundershower, tidal effect, terrestrial magnetism may cause interference pretty often. The majority of the cases where sensors are deployed in the environment, coupled with animals, fish, or deployed in underground or even undersea, are majorly affected with such kind of interference effects. Power line (50–60 Hz) frequency is another one of the major causes of interference on the sensor data in various buildings, electric grids, and substations as well as in hospitals and medical stations.

The artifact, on the other hand, refers to the noise components superimposed on the base signal. Majorly, this is due to the movement. The epidermal layer of human skin often generates the static electricity that causes such artifacts while the skin gets connected with the electrode. The signals grabbed by the IMU (inertial measurement units) while moving are majorly suffered by such noise. Often ECG and EEG signals have a resemblance with motion artifacts. Therefore, it is sometimes difficult to remove such noise from the original signal components.

1/f noise often known as flickering is characterized by its power spectrum where power is inversely proportional to the low frequencies. When current passes through semiconductor-based sensing devices, it causes a fluctuation or jittering. 1/f noise is also generated when current passes through the resistors as well.

2.1.2 Sensing and measurement

In the case of sensor data measurement procedure, the observer gets the information of the sensing entity. In major cases, the information comes from the electronic, mechanical, and electromechanical devices. In some cases, the sensing requires additional excitation of the sensing unit. For example, the optical heart bit monitoring sensor requires a constant LED illumination whose light should be passed through the blood vessels or artery. Gas sensors often need an inductor coil to be heated for functioning properly. While designing a sensor and using them, some of the crucial parameters should be observed and analyzed to understand the operational behavior of the sensing devices. The parameters, namely, resolution, precision, and sensitivity, are some of them. The sensitivity is a term that signifies the character of the sensor where a small change in the input makes a large impact on the output. Sometimes it is also considered as the ratio between the output and the input. For some sensors, the ratio is very high, which signifies a highly sensitive sensing device. We can perform some sort of calibration to manage the sensitivity of the sensor unit and the device.

Resolution, on the other hand, is a ratio that signifies the maximum measurable signal by the smallest part that can be resolved. In the case of a digital infrastructure, it is nothing but a bit change for a corresponding signal-level change. In normal case, the resolution of the sensor can be determined in terms of voltage and bit representation. Suppose a sensor can do its operation within $\pm n$ volt range, which can be converted using K-bit A/D converter. So that total combination we can found with K bit is 2^K. Therefore, the resolution R of the system can be written as:

$$R = \frac{2n}{2^K} \qquad (2.3)$$

Therefore, the smallest change that can be detected by the sensor is equal to R.

Precision is defined by the reproducibility. It signifies that how two measured data are closer to each other. It completely depends upon the range of data supplied by the sensors. If the range is narrow, the precision is high; if the range is wide, then the precision is low.

2.1.3 Calibration and error scenarios

Calibration of the sensor is a crucial issue and can be done critically. In most of the cases, the calibration has to be performed by comparing the readings with the standard instrument or some standard quantities. For example, a barometer can be calibrated with a standard sea-level pressure of 760 mm of mercury. In some cases, the ice point of water or melting point of gallium is also considered as standard value for the calibration. An instrument that is calibrated with a standard measurement can be considered as a standard instrument, and it helps us to calibrate other equipment pieces. A two-point calibration is standard when the sensors are linear. In the case of a linear system with drifting nature, the oscillation point calibration and the two-point calibration are recommended.

In a standard measurement scenario, the sensors may produce several types of errors and they also come from different sources. Some of the major types of error that the sensors may experience are random error, quantization error, dynamic error, and systematic error.

Random error is a type of error that happens in an unpredicted manner. It imposes in the baseband signal as a short-term fluctuation. In general, the deviation of the error has spread in a wider range of negative and positive values. It is expected that the mean of the deviation is zero. And if the error repeats while measures quantity remains unchanged, then the average of the measured value is considered to be the true value.

In the common case, the quantization error happens mostly for the analog sensors. When the analog signal gets converted into digital data, a sampling of the analog signal is performed by a sample and hold circuit. To present the sample with a proper digital bit, sometime ambiguity happens. If quantization error happens in the least significant bit position, then in most cases it is trivial. On the other hand, if the error rose in MSB (most significant bit) position, then the matter is serious. To avoid such quantization error, the sensor data should be adjusted with resolution and the input range of the analog-to-digital converter of the system.

A dynamic error occurs due to the dynamic characteristics of the measurement unit. Such error occurs while signal value changes so abruptly and quickly that the system cannot follow the relation between the input and the output. During dynamic error, the instantaneous output of the measurement system never reflects the input value. This problem often arises in the MEMS device of having the first order and the second order.

Systematic error mainly appears in a repeated interval. Systematic error is sometimes reported as a very low-frequency component or noise, which cannot be segregated easily from the original base signal even after repetitive measurement and averaging. Sometimes systematic error happens due to environmental parameters like the change of climate. In such cases, two identical measurement units can be used simultaneously, and the difference of the output value is useful to keep track of the systematic error. If the change in the output is the same foe in both units, then the error gate cancels out automatically.

2.2 Flow sensing and measurement technique

In the human body, the majority of the part is fluid. Therefore, flow sensing and measurement is an utmost important and crucial challenge [1]. When blood flows from the artery to the tissue, the amount of flow can be measured by the volume flow rate per unit mass of the tissue. In general, the unit of such flow is considered as mL/min. The flow of the blood through the blood vessel is roughly estimated by the size of the vessel. The velocity of the blood flow is also not uniform in general and perhaps a parabolic. The parabolic velocity profile can be realized by the following expression:

$$V(r) = V(max)\left(1 - \frac{r^2}{R^2}\right) \tag{2.4}$$

Here $V(r)$ is the velocity at a point of distance r from the center. $V(max)$ is the maximum velocity, and R is the internal radius of the blood vessel. The flow profile of the blood through the blood vessel is depicted in Fig. 2.1.

Blood flow in the tissue mainly differs in different physiological conditions. The blood flow profile in the artery sometimes locks like a convex lance to a parabolic structure. Another important aspect is the flow imaging through which the blood flow through the local tissue can be mapped and monitored. This is another important aspect of the clinical and physiological study. We can consider various types of flow meter equipment such as electromagnetic flow meter and ultrasound-based flow meter. In the next section, the working methodology of ultrasound-based flow meter has been described.

2.3 Ultrasound-based blood flow sensing

Ultrasound sensing is a classical way to measure the various parameters in tissue [2]. To realize the blood flow, the ultrasound-based measurement is possible. Ultrasound is a sound wave that is higher than 20 KHz. It propagates through tissue at a speed of 1500 m/s.

Fig. 2.2 represents the advanced sensor data monitoring through the ultrasound sensor module. In this case, the sensor module has three parts: a transmitter transducer, a receiver sensor, and a reflector. The propagation of the sound through the human body solely depends upon the transmission velocity, the impedance of the body, and the attenuation

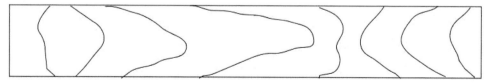

Fig. 2.1 Flow profile of the blood through the blood vessel in a single cardiac cycle.

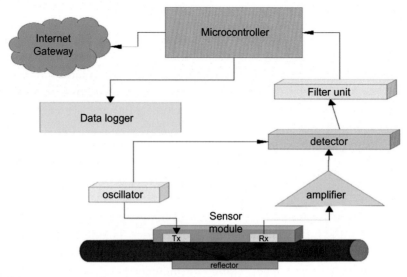

Fig. 2.2 Flow monitoring technique through blood vessels.

factor. The speeds of the sound through the hard medium like bone and muscle are always greater than that of air in STP. Body impedance and the attenuation factor are also two very crucial components that have to be taken care of. There is a direct relationship between density, impedance, attenuation, and sound speed. In general case, the higher the density, the speed will be high. The attenuation of the sound signal is also considered to be high in a high-density object. In the boundary between the media of two different densities, the sound wave reflects partially. If the sound amplitude a_i is fallen perpendicular into two varying impudence materials Z_1 and Z_2, respectively, then the resultant amplitude a_r can be obtained by:

$$a_r = \left[\frac{Z_1 - Z_2}{Z_1 + Z_2}\right] a_i \qquad (2.5)$$

To generate and receive the ultrasound, two piezoelectric zirconate titanate crystals (PZT) have to be used. When a continuous signal is applied to the transmitter signal, a specific ultrasonic signal pattern gets generated. The signal gets reflected in the reflector. Depending upon the geometry of the source and the reflector, the pattern of the wave gets determined. Finally, the signal gets received by the receiver transducer. As it is received by the *Rx* unit, the amplitude of the wave gets attenuated. To amplify the signal in such case, we have to use an amplifier. Then, the amplified signal gets detected by the detector circuit. The output of the detector circuit is then connected with a filter. The responsibility of this circuit is to truncate and block unwanted noise and the frequency component that is considered as artifacts and resonance. Finally, the signal gets fed to the microcontroller or signal processor for further analysis. The signals data might be logged in a local device, or maybe the device gets connected with some cloud service to perform analytics on the signal value.

2.4 Force-sensing measurement

Body motion and the force-sensing measurement is one of the crucial aspects in the clinical diagnosis of various diseases, like Parkinson's diseases, osteoarthritis, rheumatoid arthritis, and many more. The gait posture and motion sensing are perhaps the most common terms to identify the stages in the severity of such diseases [3]. In our muscle, several kinds of motion have been generated. Generally, three classes of muscle can be considered: skeletal muscle, cardiac muscle, and smooth muscle. The contraction speed, length change, and tension are depicted in Fig. 2.3.

Skeletal muscle contraction is quick, and the tension is also large. Cardiac muscle's contraction and tension, on the other hand, are very small. Smooth muscles sometimes contract slowly but with a large tension. The muscle acts as an actuator in the human body. A large muscle in the body often generates a large tension but the contraction is not so quick because of the larger mechanical load. Movement and displacement can be measured using various types of sensors, like potentiometer, magnetic, and capacitive sensors. Electrogoniometer is one of the potentiometer-type sensors that can be used to track the movement of the muscle and joint. This consists of a potentiometer or

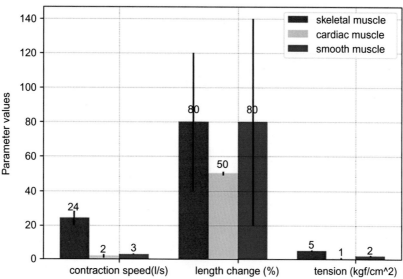

Fig. 2.3 Contraction speed, length change, and tension comparison of skeletal, cardiac, and smooth muscles.

conductive rubber band. The movement of join comprises several degrees of freedom. In order to get a precise measurement, three rotational potentiometers or conductive band can be used for three different axes. For the motion study of knee and hip joint, this technique is highly popular. Fig. 2.4A shows that the electrogoniometer systems consist of a rotational potentiometer and conductive rubber band, respectively.

Another crucial noncontact method of rotation and locomotion measurement technique is the photographic technique shown in Fig. 2.4B. This can be done by processing the available video information. One of the popular kinds of motion measurement technique is known as gait analysis. In this technique, the subject moves through a gait path and his motion gets recorded in the video camera. Some retro–reflector marker is placed in various joint points like knee, wrist, shoulder, etc. The marker illuminates an infrared signal. The position of the marker is tracked using an intelligent tracking mechanism in this case. In this way, the gait analysis can be done with a minimum error rate of 0.1%. The light-emitting diode can also be used as a marker to point the position more

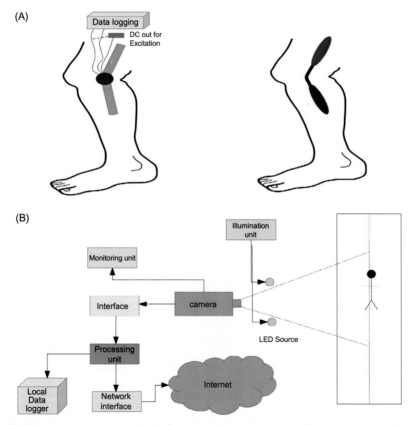

Fig. 2.4 (A) Electrogoniometer systems: left with consists of a rotational potentiometer right side with conductive rubber band and (B) motion measurement unit for gait analysis.

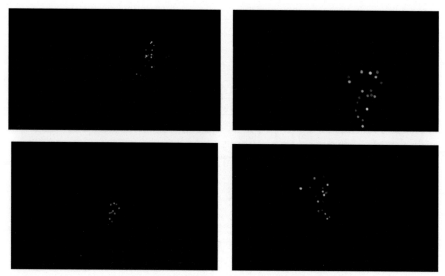

Fig. 2.5 3D representation of the gait of diplegia patients of different stages.

accurately as well. A motion disorder like diplegia, which is a class of cerebral palsy, can be diagnosed through the gait analysis pretty often [4,5]. To acquire the data, a set of electromyographic sensors and the optoelectronic marker have to be placed over the patient skin. These sensors grab the patient gait kinematics and muscular activities. Fig. 2.5 presents the 3D representation of the gait of diplegia patients.

2.5 Foot force measurement using smart shoe

In many of the applications such as gait analysis, the foot force measurement is one of the important techniques through which the mobility anomaly of the human body can be detected [6]. This data is useful to detect the patient of Parkinson's disease and arthritis. Not only that the health condition of the athletes can also be monitored by using such shoes. To design a smart shoe, majorly in-shoe pressure sensor can be used. During the level walking, the sum of the Z-coordinate velocity of the center of mass becomes zero; therefore, the momentum in Z-direction can be expressed as:

$$S_z = \triangle p_z = m \triangle v_z \qquad (2.6)$$

In this case, S_z is the impulse and $\triangle p_z$ is the momentum change. Therefore, the final equation of the momentum along with Z-direction yields

$$\int_{hsl_i}^{hsr_j} (FZ_l + FZ_r + W)dt = m \int_{hsr_l}^{hsr_j} a_z dt \qquad (2.7)$$

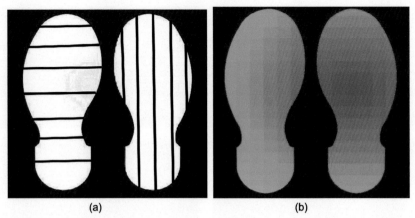

Fig. 2.6 (A) Vertical and horizontal placement of piezoresistive sensing plate for foot pressure sensing and (B) pressure map for the force sensed by the piezoresistive sensors.

To develop, the hardware carbon-embedded piezoresistive material can be used. This thin sensing plate gets sandwiched inside the layers under the sole. The electrode can be placed in the horizontal and vertical directions, as shown in Fig. 2.6A. However, Fig. 2.6B shows the pressure map obtained from the foot sensors.

2.6 ECG sensing and measurement

The majority of the medical sensors nowadays can be categorized into two parts, namely, in-body sensors and on-body sensors. The main goal of the body sensors is to acquire the parameters of the inner part of the body. We can consider an in-body sensor network to monitor and track the diseases. Such a sensor uses 802.15.6-2012 standard to communicate with the network. Such sensors mainly communicate in the Medical Implant Communication Service (MICS) band. The propagation medium for such sensor networks is the body tissue itself. This leads to a crucial signal attenuation problem. On the one hand, the data rate for the communication is not so much low in this case. On the other hand, the on-body sensor networks help us to find out the measurement of the body parameters directly from over the skin. One of the popular examples of such sensor devices is ECG sensors. ECG is a recording procedure of the electrical activity of the heart. The recording is mainly done by using some electrodes placed above the skin. Nowadays, the 12-lead ECG is highly used as multichannel ECG system. These classes of the system are mainly used to measure the short-term ECG signals within a resting position. For such ECG devices, the capacity is limited to several minutes. The long-term ECG monitoring systems, on the other hand, use a Holter monitoring [7,8]. With the recent development of the advanced transducers and the communication equipments, high-quality ECG data can be generated with minimum error.

2.6.1 Electrocardiogram systems

The ECG or electrocardiogram is a famous device or tool that is used to diagnose the condition of the heart. ECG can provide support and evidence in diagnosis [9,10]. It is an effective tool for diagnosis and management. The major parameters that are to be considered by ECG to diagnose are the abnormal cardiac rhythms. The simple ECG signal can be easily analyzed with a pattern reorganization method. Nevertheless, some basic rules can also be applied to easily understand the ECG signal pattern. In general, the contraction of any kind of muscle generates electric charges. These charges are often called depolarization. These changes can be easily detected with the electrode attached to the body surface. When the patient is the full relaxing position, all muscle contraction gets reduced and the contraction related to heart is getting recorded. If we consider the electrical point of view, the human heart chamber consists of two chambers because two atria contract together, and after that, two ventricles also contract together.

There are some specialized areas of right atrium that are responsible for the electrical discharge for each cardiac cycle often called sinoatrial nodes or SA nodes. When the depolarization occurs, it spreads through the muscle fiber and the propagation delay is also happening. The depolarization happens in another zone called the atrioventricular node or AV nodes. The depolarization pulse generally travels pretty rapidly within the area called the bundle of His. Within the mass of ventricular muscle, the conduction flows slowly through Purkinje fibers. The rhythm of the hear often called sinus rhythm begins with the electric activation within the SA nodes. The rhythm is the component that controls the heart's activation sequence. Fig. 2.7 shows a sample ECG wave. The ECG wave has several components.

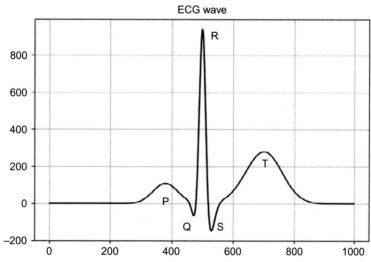

Fig. 2.7 A sample ECG waveform.

The electrical changes that accompany the atria are small. The contraction of the atria is signified by point P. In Fig. 2.7, P, Q, R, S, and T signifies the deflection of the wave. QRS part altogether is the complex component of the wave. The part beginning with S and ending with T is called ST segment. In some cases, an additional wave also can be seen after the T segment is often called U wave. The QRS complex part is the most vital in the ECG wave. The first downward deflection is the Q wave; after that, an upward deflection there is called R wave. The deflection underneath the baseline is that the following R wave is called the S wave.

In the early days, the ECG machine records the electrical activity by drawing on a moving paper. The standard rate of ECG recording is 25 mm/s in that case. Here, the PR interval has to be measured form P wave to the beginning of the QRS complex. This signifies the time taken for the excitation of the SA node through atrial muscle and AV node. The short length of the PR interval signifies that the atria have been depolarized from closed to the AV node. On the other hand, the length of the QRS complex shows how long the excitation is taking to spread. In a normal case, it is of 120 ms in any abnormal condition, which takes a longer time in comparison with others.

2.6.2 Calibration and lead

The P wave QRS complex and T waves generally provide a limited amount of information through which the machine has to be calibrated accurately. The standard 1 millivolt (mV) must move the pick of the signal 1 cm vertically. The calibration signal must be included with every record.

The lead-in ECG signal is perhaps one of the confusing terms. The term in general sometimes implies the wires that are connected with the ECG. But it is the electrical picture map of the heart. Majorly, the electrical signal can be sensed and detected from the surface of the heart through the electrode. Each of the electrodes joins in each limb, and six in the front of the chest. The system detects the signals and compares the electrical activity and hence maps the signals of different electrodes. The picture that is obtained from the different lead is different. For example, lead 1 gives the signal that is detected by the electrode set in the left and the right arm. Each electrode gives a specific lead pattern. The ECG that has been made from the 12 different characteristics view of the heart is 12-lead ECG. We can interpret the ECG signal pretty easily if we remember the direction of the lead from various corners. The six standard leads perform recording from the electrodes attached to the left and right limbs. This can be thought of as looking at the heart in a vertical plane. Fig. 2.8 depicts the ECG lead pattern recorded from six different leads.

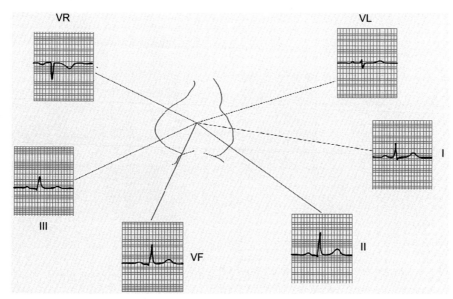

Fig. 2.8 ECG lead pattern recorded from six different leads.

2.6.3 Real-time IoT-based ECG sensing application

In today's world, the use of the IoT-based ECG sensing system and devices is highly necessary. This can be frequently used for the patient who has heart diseases and has needed constant remote monitoring. In a smart home scenario, the elderly people whose health condition is not so good can be remotely observed from hospitals or medical centers. In recent pandemic situations, the utilities of such devices and applications can be done rapidly. In the case of the highly infectious disease like COVID-19 [11], it is having a high risk to the doctor and the medical staff to go and measure the heart rate directly with the close contact of the patients. Therefore, the pretty straightforward way is to engineer a device so that monitoring can be done in real time remotely not even within the same network but within a vast geographical region. In the following work, a lightweight edge-enabled IoT application has been designed and demonstrated to perform real-time heartbeat monitoring. To physically realize the device, the system can be fragmented into some components. The architecture of the system can be considered as a bottom-up layered approach wherein the bottom layer of the ECG sensing nodes are present. The sensor nodes aggregate the data in a local processing node, which can serve as a data collector. The data collector node then connects with a gateway to pass the data packet to the edge or cloud device. In the following example, we have devised a local-level edge server that can perform data gathering logging and display the data through the dashboard. The architecture of the same is depicted in Fig. 2.9.

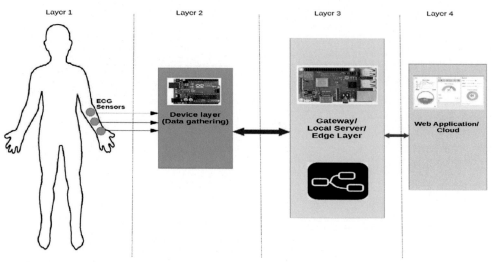

Fig. 2.9 The layered architecture of the IoT-based ECG application.

To physically realize the system, we can generally use a simple Arduino UNO as in the device layer [12]. The fundamental job of this device is to receive the analog signal and the lead-off detection signal from the ECG sensor module. In this case, typically we have used the AD8232 ECG device. This is a tiny little chip that measures the electrical activity of the heart. This is a signal conditioning board majorly used for bipotential measurement. This device can connect with three probes that can be fitted in the left or right arm. In some cases, the probe might be fitted in the left, right arm as well as in the right leg. Generally, the ECG 3 probes are accumulated in a single module format and connected with AD8232 through a 3.5-mm jack connection. There are six leads, namely, GND, Vcc, output A0, LO−, LO+, and SDN. The operating power of the device is +3.3v. The output of the device gives an analog signal, which can be feed into Arduino analog input channels. LO− and LO+ lead-off detection. This shows the impedance difference between each of the differential sensing electrodes. The SDN terminal is shut down and in general not in use. The connectivity of the AD8232 device with Arduino is illustrated in Fig. 2.10. We can upload the following code to Arduino to grab the ECG data from AD8232.

We can monitor the same data in the serial plotter of the Arduino IDE. Further, the collected data are then sent to the Raspberry Pi where the Node-Red server has been deployed. Node-Red is a lightweight platform for IoT application development. The fundamental framework of Node-Red is Node.js. This provides a drag and drop facility to create the applications. Each element in node-red is considered as the Node. To create an ECG sensing application, we can use a serial node, a function node, a chart, and a gauge node from the dashboard. The serial node receives the data available in the serial port. The port and the baud rate must be matched with Arduino serial settings. Data in

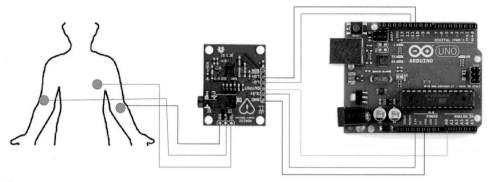

Fig. 2.10 Circuit diagram and the connectivity of AD8232 and Arduino.

the serial port, in general, delivered an ASCII string, and the splitting of the data is done based on "\n" (newline) character. The default response timeout is set to 10,000 ms. To deploy it as a real-time IoT application, the data should be fetched through a serial node of Node-Red. A serial node is the node object that can sense the data from the serial port open in the system. It is having several important properties like baud rate, data bit, parity, and the stop bit. Along with this, we can set the split input character as well as the type of the output message whether it is ASCII string or binary buffer. The split input is mainly used to separate a message sequence into two or more different message sets. Fig. 2.11A and B shows the output dashboard screen of node-red and the flow window respectively. In this case, the node-red is deployed in the local device only. To access the application globally, we can deploy the node-red instance in the cloud environment as well. In that case, the best choice is to deploy node-red instance as an app in the IBM Watson cloud. To do so first of all, we have to lunch a node-red app. Once a node-red app is created, we have to give a specific name so that we can access that app using that name. After that, we have to create a cloud foundry app. This should include a node.js framework. After launching cloud foundry, a service app must be created.

Cloud-ant app is a service that controls the functionality of the node-red app, and it makes a connection between the node.js framework and Node-Red. Now the service must be deployed. As it gets deployed, a continuous service will also start. The continuous service in this case will ask for an API key. Therefore, we have to generate a new API key and select a server region where the continuous service should deploy. After setting up the key, the continuous service will start its deployment. We must wait for 5–15 min depending upon the cloud server load to fully start the node-red service. Fig. 2.12A–C shows the different components in IBM cloud node-red setup.

As the node-red app has been successfully deployed in the cloud, now it's time to send the ECG data to the cloud. There are several ways to do that. One technique is to deploy an IoT app and register the device in Watson IoT. The second and simplest technique is to use the node-red app in transmitter-receiver mode. In this case, we have to run 2 node-red instances. The first instance should run locally in PC or raspberry pi, while

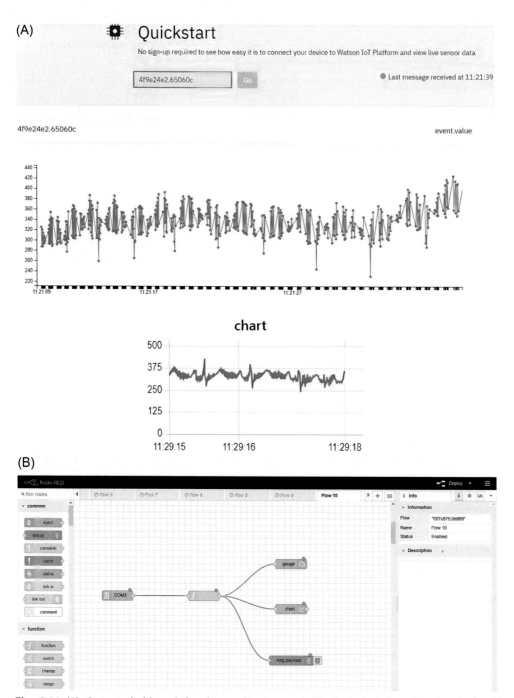

Fig. 2.11 (A) Output dashboard for the application and (B) Node-Red flow for the deployed application.

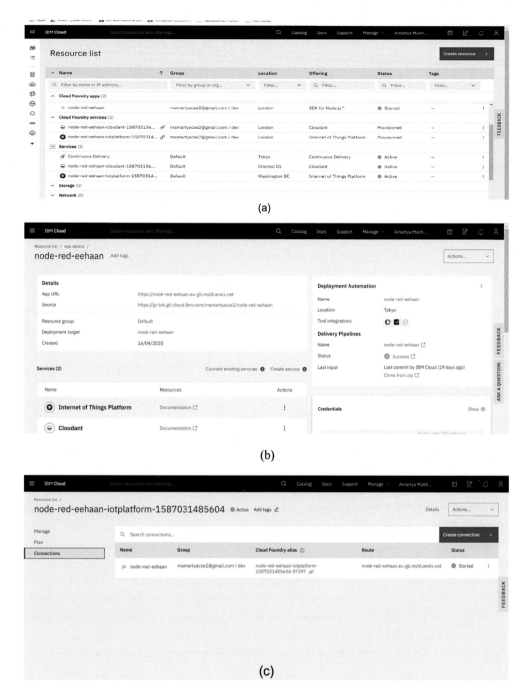

Fig. 2.12 (A) Resource list for node-red in IBM cloud platform, (B) application details with deployment status, and (C) connection details for Node-red application deployed in IBM cloud.

the second instance should run in the IBM Watson cloud. To send and receive messages in that case, we have to use MQTT protocol [13]. In such a scenario, the use of the MQTT broker is also vital. To forward the message from one device to cloud, we should not use a local MQTT broker; rather, we have to use a public broker. One of the popular MQTT brokers, in that case, we can use is test.mosquitto.org. This is a publicly available broker to test the MQTT message relay between the publisher and the subscriber. The test Mosquitto server listens to various ports; in general, 1883 port is dedicated for unencrypted message transfer. Port 8883 is for message transfer in encrypted mode. Port 8080 and 8081 are dedicated to web-socket mode (unencrypted, unencrypted, respectively). Port 8884 is also present, and in this case, the client certificate is required.

2.6.4 Heart disease prediction mechanism

The heart disease detection and prediction mechanism is a crucial approach to get an overview of whether a person has a chance to evolve the critical heart disease. Early detection and the prediction mechanism help to survive the potential heart patient's life. There are numerous mechanism that can be designed and modeled to detect and predict the chance of heart disease for a particular person. Amin B. et al. proposed a machine learning model that predicts the heart disease by considering features such as age, sex, chest pain location, pain provoked by exertion, after rest relieve, blood pressure after rest, smoking habit, family history of heart diseases, and maximum heart rate achieved. Fig. 2.13 illustrates the dataset and corresponding crucial features. The feature data, in this case, have been collected from the UC Irvine Machine Learning Repository. To design the model, the core components used are Kernel principal component analysis (KPCA) and multilayer perceptron classifier (MPC). A pipelining approach has been performed to ensure the coordination between two core components.

A Principle component analysis is a tool that is used for dimension reduction operation [14]. It allows the dataset to reduce the dimension without losing much of the information. The dimension reduction process involves finding out orthogonal linear combinations of the original variable with a large variance. Mainly, the first principal component captures most of the variance in the data. Second is orthogonal to the first principal component. The job of it is to capture the remaining variance. Thus, the main objective of this technique is to convert a set of observations of the correlated variable into the linearly uncorrelated variable. The KPCA, on the other hand, is a generalized approach of PCA that uses the kernel method. Given a set of input vectors x_t ($t = 1 \ldots l$ and $\sum_{t=1}^{l} x_t = 0$), the dimension of each is considered as m; therefore,

$$x_t = \{x_t(1), x_t(2) \ldots x_t(m)\}^T \text{ where } m < l \tag{2.8}$$

In the case of KPCA, the input vector x_t into higher-dimensional feature space $\phi(x_t)$. After that to compute the linear PCA in $\phi(x_t)$. Linear PCA can be mapped to the

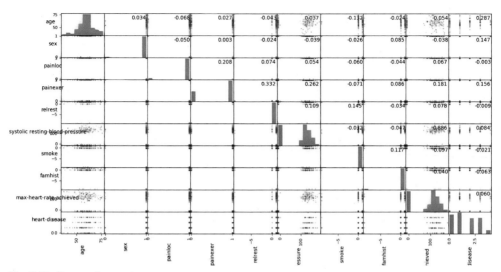

Fig. 2.13 Dataset features.

nonlinear PCA in x_t. The dimension is supposed to be larger than the training samples l in this case. This can be solved as eigenvalue problem as

$$L_i u_i = C \cdot u_i, \text{where } i = 1 \ldots l \tag{2.9}$$

Here, $C = \frac{1}{l} \sum_{t=1}^{l} \phi(x_t)\phi(x_t)^T$ is considered as a covariance matrix of $\phi(x_t)$. L_i is an eigenvector having nonzero value. $C \cdot u_i$ is the eigenvector here. Therefore, Eq. (2.2) can be converted into an eigenvalue problem as $L_i u_i = K \cdot u_i$, where K is the kernel matrix of dimension $l \times l$. The value of K can be computed, which is equal to the inner product of two vectors x_i and x_j, which is in the higher-dimensional feature space. We can consider α_i corresponds to eigenvector K that satisfy $u_i = \sum_{j=1}^{l} \alpha_i(j)\phi(x_j) \ldots$, where $\alpha_i(j), j = 1, \ldots l$. is the component of α_i. Each α_i value, therefore, is normalized using the corresponding eigenvalue by considering the following equation:

$$\alpha_i = \frac{\alpha_i}{\sqrt{L_i}} \tag{2.10}$$

Now form the estimated α_i; the principle component $s_t(i)$ can be derived for x_t as follows:

$$s_t(i) = u_i^T \phi(x_i) = \sum_{j=1}^{l} \alpha_i(j)K(x_j, x_t), \quad i = 1, 2, \ldots l \tag{2.11}$$

The multilayer perceptron (MLP) [15], on the other hand, is an essential component that can be considered as a feed-forward artificial neural network. In general, a multilayer Perceptron may consist of three or more than three layers. It comprises the input layer, output layer, and one or more than one hidden layer. The main ingredient of the multilayer Perceptron is the activation function. The activation function is a linear function that maps the weighted input toward each output neuron. The activation function f is also known as the sequencing function. It keeps the output of the neuron at a certain level like an actual neuron. There are various types of activation function, which can be considered. Among them, the most common is the sigmoid function, which can be expressed as follows:

$$y_i = \frac{1}{1 + e^{-z_i}} = f(z_i) \tag{2.12}$$

Another very useful activation function is also a kind of S-shaped sigmoid function, can be expressed as

$$y_i = \frac{1 + \tan h(z_i)}{2} = \frac{1}{1 - e^{-2z_i}} \tag{2.13}$$

In this case, the KPCA and the MLP are considered within a pipeline. The KPCA itself is used as a transformer in the pipeline. The different components and the clusters can be managed and transformed by using KPCA very easily. KPCA helps to organize data in a meaningful form. Otherwise, the data may be unorganized. MLP, on the other hand, serves as an estimator in the pipeline. The main job of the Perceptron here is to assign the weight values into different neurons. The weights have therefore been adjusted in the neuron by observing the training data. Each time one row of data has been inputted into the Perceptron, and the result is compared to the expected output. From that, the stochastic gradient error has been computed. The error value has finally been backpropagated within one neuron at a time so that neurons get the error values and try to minimize the error by adjusting the weight values. The implemented algorithm is illustrated in Algorithm 1.

Algorithm 1: Multilayer perceptron predictor

```
1. Input: age, sex, painloc, painexer, relrest, systolic resting-
   blood-pressure, smoke, famhist, max-heart-rate-achieved,
   heart-disease
2. Output: predicted (true/false), precision, recall,
   F1-score
Start
3. imputer ← Imputer (missing_values=-9,
   strategy='most_frequent', axis=0)
4. imputer.fit ← imputer.transform() // replace with new
   data
5. X_train, X_test, Y_Train, Y_Test ←
   train_test_split(X_resampled,
   y_resampled,
   test_size=0.25) // performing split for training and
   testing dataset
```

```
 6. new ← train_test_split(X, Y, test_size=0.25)
 7. X_test ← new[1]
 8. Y_Test ← new[3] // using of actual data for tests
 9. sc_X ← StandardScaler() // perform feature scaling
10. X_train ← sc_X.fit_transform(X_train)
11. X_test ← sc_X.transform(X_test)
12. clf ← MLPClassifier(solver='lbfgs',
    learning_rate='constant', activation='tanh')
13. kernel ← KernelPCA()
14. pipeline ← make_pipeline(kernel, clf)
15. pipeline.fit(X_train, Y_Train) // pipelining with PCA and
    MLP
16. y_pred ← pipeline.predict(X_test)
17. report ← metrics.classification_report(Y_Test, y_pred)
18. Output ← report
19. answer ← answer.reshape(1,-1)
20. answer ← sc_X.transform(answer)
21. Output ← pipeline.predict(answer)
End
```

The results that are obtained from the prediction value are a kind of binary classifications that show whether there is a chance of heart disease or not with 1 or 0 values. Table 2.1 illustrates the input and the result for positive and the negative cases.

The precision, recall, and the F1 score corresponding to the prediction of positive or negative symptoms have been plotted in Fig. 2.14A–C. Here, the precision value implies how precise or accurate the model to predict the true-positive cases precision metric is the best way to determine the model functionality. Mathematically, precision P can be considered as

Table 2.1 Input parameters and the predicted output values for two patients with positive and the negative cases.

Parameter	Value for patient 1	Value for patient 2
Age	66	54
Sex	1	0
Painloc	0	1
Painexer	1	0
Relrest	0	0
Systolic resting-blood-pressure	180	124
Smoke	1	0
Family hist	1	1
Max–heart–rate–achieved	210	140
Predicted result	1 (yes)	0 (no)

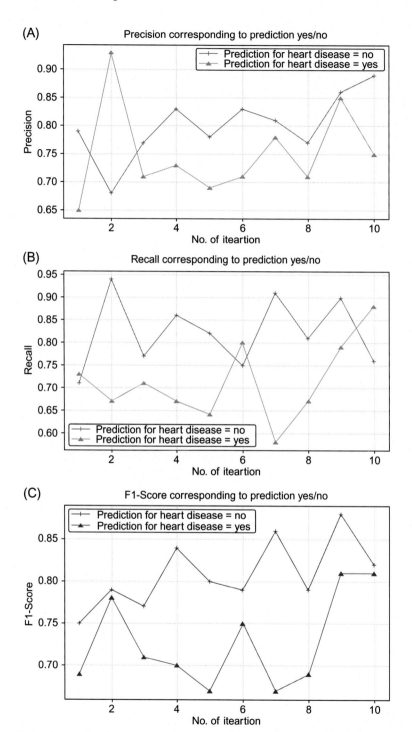

Fig. 2.14 (A) Precision value for corresponding to predicted cases, (B) recall value corresponding to predicted cases, and (C) F1-score corresponding to predicted cases.

$$P = \frac{Tp}{Tp + Fp} \qquad (2.14)$$

This implies the ratio between true-positive value (Tp) and the total predicted positive values, which are a true positive plus false positive (Fp). Precision is a good metric when the false-positive cost is high. In this case as Fig. 2.14A shows the precision, the value is majorly increasing in the heart diseases prediction of "no" at the 10th iteration.

It has been observed the highest precision value, which is near about 0.9. On the other hand, the precision of "yes" prediction is highest in second iteration. In Fig. 2.14B, the recall value has been presented. The recall value gives us the ratio between true positive and the total actual positive. Hence, recall R can be computed as

$$R = \frac{Tp}{Tp + Fn} \qquad (2.15)$$

The recall value is useful when we have to select the best model and the high cost is associated with false-negative values. The recall corresponding to "no" prediction reaches up to 0.95 in 2nd iteration, whereas the recall value for "yes" prediction goes up to 0.88 in 10th iteration. However, in 7th iteration, the recall for the "yes" prediction reaches its lower benchmark, which is below 60%. To make a trade-off between precision and recall value, the F_1 score has to be computed. It can also be considered as the measure of the test accuracy. It is the harmonic mean of the precision and the recall value. If the value reaches 1, it suggests the perfect precision and the recall. Mathematically, F_1 score can be expressed as

$$F_1 = 2 \times \frac{P \times R}{P + R} \qquad (2.16)$$

Fig. 2.14C depicts the F_1 score variation for alliteration from the figure; it is also clear that F_1 score for the "no" prediction is higher than for the "yes" prediction. It means the model is best for predicting the true-negative occurrences. In iteration 9, the score is reached almost 0.98, which is reported to be the highest benchmark for the F_1 score in this case.

2.7 EEG fundamentals

Electroencephalography of the often-called EEG is a component that plays a pivotal role in the diagnosis of neuron diseases as well as brain study [16,17]. The extensive research and the study of brain signals offer a better understanding of the dynamics of the brain. The major category of epilepsy diseases is identified and classified using EEG. Not only that another major application of that is the brain-computer interface (BCI). Here, the signal helps recover the sensory and motor functions of the patient who has a major motor disability. The fundamental concept of the EEG signal is the recording of the electrical activity that happens in our brain. It is experimentally proved that how brains state changes over time and the scientists over the year study the functionality of the brain by capturing EEG signals. The study of the brain using EEG signal is perhaps the most crucial tool to detect the crucial neurological problems.

The main component of the EEG test involves EEG signal capture. This can be done by placing the proper number of disk-like electrodes. Each electrode is generally connected with an amplifier. The amplifier is further connected with the EEG recording device. The electrode should be powerful enough to detect tiny electrical charges that appear within the brain during the activity of the brain. In case of recording of the EEG signal from 1 to 256, probes can be placed overhead in parallel to get more accurate signals. Such a technique is often known as multichannel EEG recording. There are two major kinds of EEG devices that one can use: (i) scalp EEG and (ii) incremental EEG. For scalp, EEG small electrode has to be placed on the scalp. Such EEG must ensure a good mechanical as well as an electrical contact. On the other hand, for incremental EEG, special electrode has to be placed underneath the scalp by performing surgery. Fig. 2.15A illustrates the electrode deployment position, and Fig. 2.15B shows the EEG signal pattern.

In general, the amounts of charges detected by the electrode are very nominal and they are the result of the activity within the brain. The charges are further amplified and finally represented on the screen as a graphical form. Typically, the amplitude of the EEG signal varies from 1 to 100 mV for adults. Due to the nonuniform distribution of the cerebellar cortex, the signal produced by the recording electrode may vary. The placement of the electrode is therefore a crucial challenge as the different lobe of the brain is responsible for different activity processing. However, the scalp electrode localization is an international standard method that is based on the international 10–20 system. 10 and 20 in this case represent the actual distances from the neighbor electrode in the order of percentage (which is either 10% or 20%) of the left-right or back and front distance within the skull. The two general position are *nasion* and *inon*. The first position signifies the position between the forehead and nose. Inon is a bony part at the base of the skull. Each location of the electrode can be identified as different letters. The letters F, T, C, P, and O represent frontal, temporal, central, parietal, and occipital, respectively. The placement of the electrode also refers to as montage. There are several montages that are available like bipolar montage: In this case, one pair of electrodes has been applied and makes up a channel. Reference montage: here each channel represents the difference between certain electrodes and a specifically designated electrode.

2.8 EEG signal analysis and classification

In this section, we are presenting the fundamentals of emotion analysis using the EEG signal. It is true that by seeing the EEG data directly the human mind state identification is quite challenging. In order to understand the state of mind, in this case a support vector

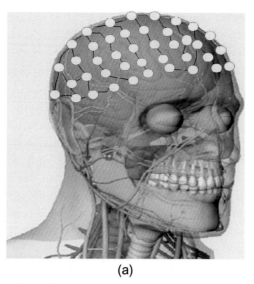

(a)

(b)

Fig. 2.15 (A) EEG electrode deployment positions and (B) monitored EEG signal pattern.

machine-based classifier is used. The classifier has been trained with the DEAP dataset. The state of the mind in this case is determined using two major parameters, namely, valence and arousal. These two parameters further are used to determine the state of mind. The valence and arousal are the main components of the circumplex model [18,19]. The main objective of this model is to realize the emotions into a circular two-dimensional space. In this case, arousal and valence signify two dimensions. These two parameters in this case represent the vertical and horizontal axis, respectively. Here, valence indicates the degree of pleasantness, and the arousal signifies the intensity of the perception. This circumplex model affects the provided insight into the neurophysiology of different neural disorders. It also offers a widespread theoretical framework understanding the widespread logic of mood anxiety disorders. Further, the dimensional approach is a statistical tool that is used in neuroimaging and neurobiological studies. From the large group of scientific studies, it was proved that the mesolimbic dopamine system has a significant role in processing reward and pleasure. The major role of the mesolimbic system in reward has perhaps been validated and demonstrated in both humans and animals. It has also been studied that not only for pleasure and reward the mesolimbic system is also responsible for drug addiction. Numerous studies have also been reported about the functionality of the mesolimbic system, which is associated with the positive and the negative emotions. Thus, the circumplex model maps those positive and the negative emotions into a circularly designed framework. The fundamental idea of the circumplex model is depicted in Fig. 2.16.

In the next segment, we have discussed an analytical framework, which is a combination of EEG signal with a data logging facility. The logged data is then reduced in dimension, and the artificial intelligence algorithm has been applied for further analytics. The block diagram of the overall sensing and processing mechanism is depicted in Fig. 2.17.

The data set that has been used is a popular dataset, and it comprises two parts. The first part consists of online self-assessment ratings with 120 one-minute extracted music and some set of videos that are rated by 15 volunteers based on their arousal, valence, and dominance feature. Second, the participant ratings, physiological recordings, and face video of an experiment have been logged from 32 volunteers who have watched a subset of 40 of the above music videos. EEG and physiological signals were recorded, and each participant also rated the videos. The labels are extracted into separated .dat file. Each channel has been stored in row-wise fashion with time in the column for each participant. The wavelet transformation has been used to decompose the data. The data here decomposed five distinct sets of features. These features are based on the frequency of the brain wave. The spectral analysis has been made from each time series by considering the classical frequency bands shown in Table 2.2.

Although there are a total of seven decomposed values that get generated. Two of them got neglected. The frequency of 0–0.5 Hz and the frequency of 50 Hz are to remove artifacts and to neglect the power line interruption, respectively.

The classification and the prediction have been done based on the support vector machine classification technique. To get the most accurate prediction, the tuning of the hyperparameters is highly required. The major parameters in such a case are the selection of the kernel value, choice of the penalty parameter C, and the γ value for choosing the nonlinear hyperplane. A fivefold cross-validation mechanism has been engineered to get a more accurate result. Table 2.3 depicts the hyperparameter values used in the experiment.

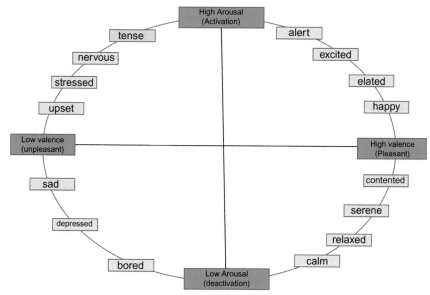

Fig. 2.16 Graphical representation of the circumplex model of human emotion. The horizontal axis represents valence, and the vertical axis represents arousal.

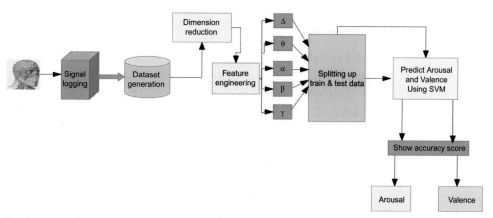

Fig. 2.17 EEG data sensing and the processing ecosystem.

Table 2.2 The classical frequency bands for spectral analysis.

Classical frequency	Frequency band (Hz)
Δ-wave (delta)	<4
θ-wave (theta)	4–7
α-wave (alpha)	8–15
β-wave (beta)	16–31
γ-wave (gamma)	32

Table 2.3 Used hyperparameter values.

Hyperparameter	Value
C	0.1–1.0
Kernel	RBF
γ	0.001–0.01

Fig. 2.18A and B depict the predicted accuracy for the varying number of C and gamma.

From Fig. 2.18A, it is clear that the increasing value of C increases the prediction accuracy. The nature of the curve for both valence and arousal is sigmoid in both cases. The penalty parameter C describes how much we can bypass the misclassification of each training set. The value of C should be always nonnegative, and a large value of C of the optimizer itself chooses a smaller margin of the hyperplane. Thus, the hyperplane performs a better job of getting training samples classified correctly. Thereby, the prediction accuracy increases. Conversely, the lower value of C forces optimizer to look for a larger margin of separation hyperplane although it misclassifies the majority of the training samples. Sometimes for some little change of C value results in misclassification although the points are linearly separated.

Further in Fig. 2.18B, it is observed a parabolic increasing curve of the prediction accuracy concerning the increasing gamma value. This parameter signifies the influence of the training example or in some case reachability. The lower value signifies "far away," and the higher value signifies "closer." The gamma parameter is also considered as inverse of the standard deviation of RBF kernel or sometimes signifies as the inverse of the radius of the influence of the sample for which the support vector has been chosen. If the value of gamma is too large, the area of influence of the support vector-only includes itself. In such a case, the higher value of regularization or penalty parameter doesn't prevent overfitting. If the value of the gamma is too small, the machine in such cases is unable to understand the shape of the data.

2.9 Ultrasound sensing for tissues and fetal growth observation

Medical ultrasound is also known as diagnostic sonography. This is a popular imaging technique and also a therapeutic application of ultrasound system. Majorly, the USG helps to build up the body's internal structure, tissues, and tendons, and also uses for the fetal activity and the growth and perinatal outcome during pregnancy. Practice of such mechanism is also called obstetric ultrasound. In normal case, the frequency of ultrasound is beyond human audible range, which is >20,000 Hz. The activity has to be performed by deploying active sensor probe that emits that ultrasound. It uses the reflection property where the sound signals get reflected in various tissues and tendons. There are

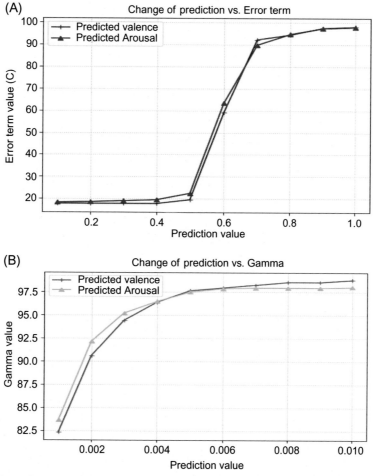

Fig. 2.18 (A) Penalty parameter value (C) vs the predicted accuracy value for valence and arousal. (B) γ (kernel coefficient) vs predicted accuracy value for valence and arousal. The number of iteration = 10.

various types of ultrasound images that can be obtained while performing it. The most popular types of image are B-mode image shown in Fig. 2.19A and B.

The effective ultrasound imaging of the superficial structure of human body can be done with a frequency range of 7.5–19 MHz. The body parts like muscle, tendons, breast, testis, and thyroid and parathyroid glands can be captured using this frequency value. On the other hand, the tissues present in the deep of human body like kidney, liver, uterus, and heart can generally be observed by using a frequency band range of up to 1–6.5 MHz. In order to get precise image data, the transducers are generally placed outside of the body by touching the skin. However in many cases, the specialized allocations of the transducers are required to get more accurate image of the internal part of the body.

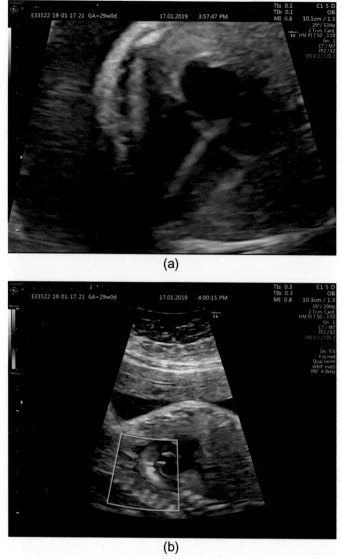

Fig. 2.19 (A) General B-mode ultrasound images for fetal anomaly detection. (B) Fetal heart position and the condition monitoring using B-mode ultrasound with Doppler measurement color coding.

Endorectal and endovaginal transducers are the popular example of such systems so that the internal activity of the organs like rectum and vagina can be observed easily.

There are wider applications of the ultrasound imaging that is present in medical system. In anesthesiology, the position and the placement of the needle can be observed during local anesthesia of the sensitive organs like nerves. For the vascular medicine treatment, B-mode UAG with Doppler flow measurement is widely used for venous and arterial disease. Intravascular ultrasound is highly preferred to identify blood clots. Generally, thin ultrasound probe has to be used in this purpose. In the field of emergency medicine, ultrasound technique is highly usable. The cases like pulmonary disorders, massive breathlessness, and pericardial tampade after trauma and also kidney and gallstones they are highly relevant. Another major study that can be done is the field of gastroenterology and colon-rectal surgery. In order to detect inflammation in appendicitis, the ultrasound is the best choice. In some cases to avoid the tomography scan, the best alternative is considered as ultrasound. The USG of liver and kidney tumors is also even helpful for the identification, detection, and characterization.

2.9.1 Obstetrical sonography

This is the most common use of the USG system done during human pregnancy. This technology was developed in the late 1960 by Ian Donald. This method is highly effective to monitor and analyze the growth and the state of the fetus. Obstetrical ultrasound is majorly used to detect the date of pregnancy, fetal visibility confirmation, fetus location checking, placenta location identification, physical anomaly of fetus, fetus heart rate monitoring, and sex identification of the baby.

Another very crucial investigation can be performed by using ultrasound technique, which is the fetal heart rate monitoring. Among all known mechanisms, the Doppler ultrasound is the most effective technique. Joint time frequency analysis is one of the effective approaches. In this technique, the average recording time of 15 patient is reported to be minimum of 20 min. In current day's perinatal medicine, noninvasive cardiotocographic monitoring is highly favorable for the observation of the fetal heart rate. The FHR (fetal heart rate) can be determined based on the duration interval between consecutive heart beats, which can be expressed as:

$$FHR = \frac{60000}{T_{RRi}[MSC]}[BPM]$$

Here, T_{RRi} is the expression of the neighboring heart rate observed. The measurement is mostly done by using embedded hardware-based system or with a computer system interface. The ultrasound transducer can be placed on mother belly and can be connected with an amplification unit. The unit is then mostly interfaced directly with the software-enabled system or a direct audio monitor. Fig. 2.20 shows an 8-month fetus heart sound captured from the ultrasound transducer deployed on mother belly surface.

The observation of 2PQRST cycle can identify six distinguished types of events: (1) atrial and ventricular wall contraction (ATC, VC), (2) mitral valve opening and contraction (MO, MC), and (3) aortic valve opening and close operation (AO, AC). It is reported that the ATC and MC reach up to 86% of the total event, whereas AO and VC are reported to be 53% and 27%, respectively. AC and MO, on the other hand, are reported to be 60% of time occurrence within the total time. The analysis shows that

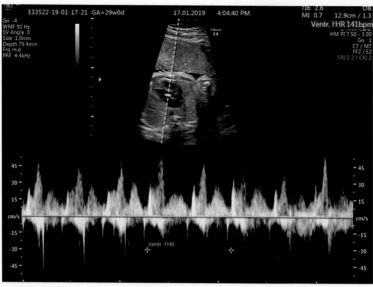

Fig. 2.20 The FHR observation of 8-month fetus sensed through the electrode applied over the belly surface. The observed FHR value was 141 bpm.

high correlation values are appearing form the wall movement and the low correlation value corresponds to valve movement. Therefore, spectrum analysis of the Doppler signal for heart beat analysis corresponding to atrium wall is utmost suitable.

2.10 Conclusion

This chapter mainly focused on sensing and data gathering mechanism for medical eco-systems. The popular sensing technology like ECG, EEG, force measurement, and USG has been discussed. The recent progress in ECG data accumulation has been emphasized and how the devices like AD8232 has been interfaced with embedded computing system and data get visualized and analyzed in cloud platform has been addressed. Further, force sensing for gait analysis and monitoring is also addressed. The state-of-the-art EEG sensing is discussed for emotion and brain mapping operations. Finally, the USG monitoring has been addressed to understand the situation of the different body parts and the fetus positioning in uterus.

References

[1] M. Hashimoto, Y. Taguchi, Kirigami-inspired electrothermal actuator for motion tracking blood flow measurement, in: *2019 International Conference on Optical MEMS and Nanophotonics (OMN)*, IEEE, 2019, pp. 152–153.
[2] N. Dey, A.S. Ashour, W.S. Mohamed, N.G. Nguyen, Acoustic wave technology, in: Acoustic Sensors for Biomedical Applications, Springer, Cham, 2019, pp. 21–31.

[3] C. Wang, D. Li, Z. Li, D. Wang, N. Dey, A. Biswas, M. Luminita, R.S. Sherratt, F. Shi, An efficient local binary pattern based plantar pressure optical sensor image classification using convolutional neural networks, Optik 185 (2019) 543–557.

[4] A. Ferrari, L. Bergamini, G. Guerzoni, S. Calderara, N. Bicocchi, G. Vitetta, C. Borghi, R. Neviani, A. Ferrari, Gait-based diplegia classification using LSMT networks, J. Healthc. Eng. 2019 (2019).

[5] L. Bergamini, S. Calderara, N. Bicocchi, A. Ferrari, G. Vitetta, Signal processing and machine learning for diplegia classification, in: International Conference on Image Analysis and Processing, Springer, Cham, 2017, pp. 97–108.

[6] Z. Li, D. Wang, N. Dey, A.S. Ashour, R.S. Sherratt, F. Shi, Plantar pressure image fusion for comfort fusion in diabetes mellitus using an improved fuzzy hidden Markov model, Biocybern. Biomed. Eng. 39 (3) (2019) 742–752.

[7] I. Nault, P. André, B. Plourde, F. Leclerc, J.-F. Sarrazin, F. Philippon, G. O'Hara, et al., Validation of a novel single lead ambulatory ECG monitor–Cardiostat™–compared to a standard ECG Holter monitoring, J. Electrocardiol. 53 (2019) 57–63.

[8] A. Appathurai, J. Jerusalin Carol, C. Raja, S.N. Kumar, A.V. Daniel, A.J.G. Malar, A.L. Fred, S. Krishnamoorthy, A study on ECG signal characterization and practical implementation of some ECG characterization techniques, Measurement 147 (2019), 106384.

[9] S. Borra, N. Dey, S. Bhattacharyya, M.S. Bouhlel (Eds.), Intelligent Decision Support Systems: Applications in Signal Processing, vol. 4, Walter de Gruyter GmbH & Co KG, 2019.

[10] N. Strodthoff, P. Wagner, T. Schaeffter, W. Samek, Deep learning for ECG analysis: benchmarks and insights from PTB-XL, arXiv preprint arXiv:2004.13701 (2020).

[11] S. Ahuja, B.K. Panigrahi, N. Dey, T. Gandhi, V. Rajinikanth, Deep Transfer Learning-based Automated Detection of COVID-19 From Lung CT Scan Slices, 2020.

[12] A. Mukherjee, N. Dey, Smart Computing with Open Source Platforms, CRC Press, 2019.

[13] A. Mukherjee, N. Dey, D. De, EdgeDrone: QoS aware MQTT middleware for mobile edge computing in opportunistic internet of drone things, Comput. Commun. 152 (2020) 93–108.

[14] H.-P. Deutsch, M.W. Beinker, Principal component analysis, in: Derivatives and Internal Models, Palgrave Macmillan, Cham, 2019, pp. 793–804.

[15] F. Shi, Y. Gaoxiang Chen, N.Y. Wang, Y. Chen, N. Dey, R. Simon Sherratt, Texture features based microscopic image classification of liver cellular granuloma using artificial neural networks, in: *2019 IEEE 8th Joint International Information Technology and Artificial Intelligence Conference (ITAIC)*, IEEE, 2019, pp. 432–439.

[16] O. Cohen, D. Doron, M. Koppel, R. Malach, D. Friedman, High performance BCI in controlling an avatar using the missing hand representation in long term amputees, in: Brain-Computer Interface Research, Springer, Cham, 2019, pp. 93–101.

[17] U.R. Acharya, Y. Hagiwara, S.N. Deshpande, S. Suren, J.E.W. Koh, O. Shu Lih, N. Arunkumar, E.J. Ciaccio, C.M. Lim, Characterization of focal EEG signals: a review, Future Gener. Comput. Syst. 91 (2019) 290–299.

[18] S.R. Jaeger, S. Spinelli, G. Ares, E. Monteleone, Linking product-elicited emotional associations and sensory perceptions through a circumplex model based on valence and arousal: five consumer studies, Food Res. Int. 109 (2018) 626–640.

[19] P. Ozel, A. Akan, B. Yilmaz, Emotion detection using multivariate synchrosqueezing transform via 2D circumplex model, in: *2018 Medical Technologies National Congress (TIPTEKNO)*, IEEE, 2018, pp. 1–4.

CHAPTER 3

Medical signal processing

Contents

3.1. Overview

Features are characteristics of a particular phenomenon; the features dictate the very way of how the prediction procedure will be modeled. The fundamental constituents of any feature are the data or signals. Biological signals are the reflection of collected action potentials of subdermal tissues of a living being. Its presence signifies the ionic and electrical activities of the muscular and the neural cells in a synchronized manner. Hence like any signals, the Biomedical Signals are also processed and important features are extracted for the very purpose of prediction. Temporal data [1] are represented by a state in time. An example of temporal data will be a collected time series data such as an EEG or an ECG signal. Data mining is the process of extracting important information from a collection of historical data. Extracting information from temporal data requires a bit of a challenge and some of the popular approaches for temporal analysis are time series, state analysis, and longitudinal analysis.

49

3.2. Time series analysis

A time series [2] can be considered as a sequence of successive points or measurement in time with respect to some variable. An example of time series analysis will be the monitoring of population with time or the monthly average temperature, etc. are all examples of time series. In the case of biomedical research, the main motivation of the research is to understand how different variables influence a particular outcome or an event over time so that regression analysis can be done. One of the major uses of Time Series analysis is the prediction of various diseases that can occur in the near future.

A time series analysis of ECG data overtime has been displayed in Fig 3.1. The peaks and troughs are explicitly colored and the rejected peaks are marked with red color, which are identified with the help of time series analysis. We have data represented in either the frequency domain or the time domain. Data represented in the frequency domain resembles how to spread the data over the range of frequency or the no. of occurrence of the same values. Time domain deals with how the data changes over time. In the procedure of data collection over time, it might be the case that there is some random variation of data collected over time due to several reasons which might include the nature of the trend or any other factor. It is important to clear the random variation from the actual data. A time series is data that is collected over time. Hence a signal is discretized and the collected data over time is gathered for analysis.

3.2.1 Signal overview

In biomedical signal processing applications, we will come across two types of signals, deterministic signals and random signals. Random signals are signals that are not governed by any rule, while deterministic signals are governed by rules. A deterministic signal can be described by a set of parameters and a mathematical function. A simple representation of a signal can be in the form shown in equation.

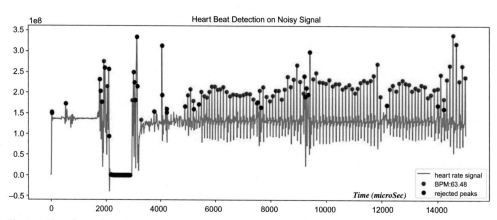

Fig 3.1 Heartbeat detection on a noisy signal.

$$x(t) = x(i + kM)$$

Here $x(i)$ is the signal, k is an integer value, and M is the time period. Discrete signals are signals consisting of sequences of quantities represented over time. The domain is restricted to a set of discrete points on which the time series is represented. An example of major usage of digital signals is the computer-aided tomography (CAT) scan where the organs of an individual are modeled in a manner with the help of digitized images from continuous X-ray plates. Magnetic resonance imaging (MRI) is another approach where strong magnetic fields and radio waves generate the images of the organs in the body. A CAT scan gives us a quick overview of the internal organs while an MRI gives a more detailed overview of the images (Fig. 3.2).

It is not that the majority of the signals found in biomedical application are discrete. There are signals that have varying quantities over time and are categorized as continuous-valued signals. Thus a continuous-valued signal can be classified as the signals whose domain is not fixed. Continuous-valued signals are usually analog signals and conversion from continuous to analog or discrete domain requires the step of passing the data through an analog-to-digital converter. Random signals fall under the domain of continuous-valued signals. Random signals can be anything: a simple thermal noise, external sound, etc. The random signals contain parameters that cannot be defined or represented with the help of a function. Hence we use various statistical tools such as probability density function to represent a random signal. Electromyogram, majorly used for diagnosis of neuromuscular disorder, collects electrical data from muscles. The collected signal data falls under the category of random signals. To test whether the signal is random or discrete there are numerous tests that are carried out. One of the popular tests is the Kendall Challis and Kitney [3] test. For a signal with n points if the no. of turning points is greater than $2*(N-2)/3$ then the signal is random. A turning point can be characterized as a change in a gradient that if the gradient changes from $-$ve to $+$ve or vice versa then there is a turning point. In order to define a turning point, three points are needed.

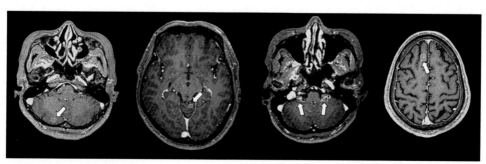

Fig 3.2 CAT scan images.

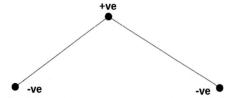

Fig 3.3 Turning point in the signal gradient.

From Fig. 3.3 we can observe three points defining a turning point in a signal gradient. If two-thirds of the points in the signals are turning points then the signal can be characterized as a random signal.

3.2.2 Some approaches

Statistical analysis and time series analysis are fundamentally used for signal processing applications.

Smoothening is one of the approaches which is applied to reduce the random variations that occur in the data. There two major groups of smoothening:

(i) Averaging
(ii) Exponential method

3.2.2.1 Moving average

In order to estimate the trend cycle of any time series data, moving average is a widely used approach before with the other complex statistical tools. This smoothing [4] approach is very simple to estimate

$$T_i = \frac{1}{p} \sum_{i=-k}^{k} y_{t+i}$$

$p = 2k + 1$, i is obtained by averaging the data in the time series with k periods values of i. Hence the findings that are nearby in the time are likely to be close in value. Hence some of the random variations in the data are removed creating a smooth trend. Hence the no. of considered instances over which a single averaging is performed is called the order, in this case the order is p. So we have a p-MA or p order moving average.

Fig. 3.4 depicts a moving average performed on ECG data. Exponential smoothening deals allotting exponential decrease weight over time. Unlike moving average where the average is estimated or weighted equally, the exponential smoothening applies weights that decrease over time in an exponential manner. The exponential smoothening is applied when there is no clear pattern in the time series data. Forecasting is done using weighted averages, but the weights decrease exponentially as observations come from further in the past—the

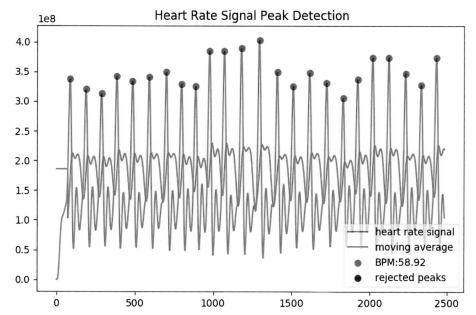

Fig 3.4 Heart rate peak detection.

smallest weights are associated with the oldest observations.A single exponential smoothening is mainly used for forecasting data with no clear trend. Forecasts are estimated with the use of weighted averages, where the weights get reduced in an exponential manner.

$$Y_{i+1} = \alpha Y_i + \alpha(1-\alpha)Y_{i-1} + \alpha(1-\alpha)^2 Y_{i-2} + \ldots$$

α can be considered as the parameter for smoothening, where $0 \leq \alpha \leq 1$. The weights decrease rate is controlled with the parameter. Y_{i+1} is the weighted average of all the estimated observation from Y_1 to Y_i.

The concept of stationary time series and difference are the two important aspects of the time series that need to be understood before we proceed with the understanding of averaging. A stationary time series can be described as a series that doesn't depend on the time it was observed, i.e. the mean and variance over time are constant for the series. In the case of a stationary time series, in the long run, there is no notable trend or pattern that can be inferred. In the concept of differencing in time series analysis, we estimate thedifference between the trend values between adjacent points in time. A simple overview of the difference in time series analysis is given in equation. One of the key uses of differencing is to stabilize the mean, remove changes, and hence the trend in the series.

$$Y_i' = Y_i - Y_{i-1}$$

3.2.2.2 Autoregressive moving average

Autoregressive models [5] or "*Autoregressive moving average*" are widely used for forecasting in the time series data where there lies some correlation between the past and future data. The approach can be considered as simple linear regression performed on the time series performed against one or more past values. An *AR model* is represented as *AR* (*p*) where *p* is called the order. An example of an autoregressive model is depicted in equation.

$$Y_i = m + \Phi_1 Y_{i-1} + \Phi_2 Y_{i-2} + \ldots + \Phi_p Y_{i-p} + \varepsilon_i$$

Here ε_i is the white noise, Y_i is a weighted moving average of the past few forecast errors, and Φ are the parameters.

3.2.2.3 ARIMA

Although exponential smoothening is a widely used model for forecasting, another important approach is the ARIMA model. ARIMA [6] stands for "*Autoregressive Integrated moving average.*" ARIMA is the combination of moving average and differencing operation with autoregression. An overview of the ARIMA is given in equation.

$$Y_i' = m + \Phi_1 Y_{i-1}' + \ldots + \Phi_p Y_{i-p}' + \theta_1 \varepsilon_{i-1} + \ldots + \theta_q \varepsilon_{i-q} + \varepsilon_i$$

Y_i' is the series difference. The above-mentioned ARIMA model can be defined as ARIMA (*p*, *d*, *q*). *d* is the degree of difference involved, *p* is the order of the autoregressive model, and *q* is the order of the averaging.

3.3. Multiscale signal processing

Medical signal processing and the preprocessing is the predominant research area. There are several processing methods and technology is being developed that is extremely useful and effective to study different disorders and diagnose and early detect diseases from signals like X-ray, ECG, PPG, tomography scan, USG, and many more. In order to properly process the medical signal information, the data preprocessing is the key part that needs to be addressed.

3.3.1 Various signal processing models

In a medical sensing application, numerous sensor nodes are deployed in various regions of the medical unit, such as the body area network, the data from several medical devices and instruments. For some elementary level data processing, noise reduction is required in each step to realizing the useful information from the data.

(A) Signal preprocessing mechanism

Signal preprocessing includes the concept of how and when the sensor data has to be taken and perform noise filtering on it. It is basically a transformation technique in which a bulk of information has to be transformed without separation of event and data classification. The preprocessing of the data therefore involves the preparation of the data for the next steps of operations. The application like control and the monitor also requires the preprocessing of the data.

(i) **Data sampling**: this is a standard mechanism. There are two types of sampling that are mainly used, namely fixed sampling and variable rate sampling [7,8]. Some other mechanisms like adaptive sampling and compressed sensing mechanisms are also popular to perform data sampling. In the case of fixed rate sampling, the data has to be sampled within the fixed frequency and the data rate. In the majority of the case, the Nyquist sampling mechanism has to be followed by the fixed rate sampling systems. Based on the hardware support the sampling power may be different in this case. Variable rate sampling, on the other hand, is more complex and a dedicated circuit has to be designed for such sampling. The main ingredient for such sampling consists of a clock generator, sampling circuits, and multiplexers. The clock produces the best sampling frequency and sends it to sample and hold circuits. It also matches with the signal coming from MCU and routed toward multiplexers. The quantization of the signal is also done thereafter. The adaptive sampling mechanism is mainly based on variable rate sampling. In this mechanism, the sampling rate gets changed based on the variable rate sampling mechanism. The sampling rate change mainly happens based on the data received from the sensors. It is a more practical approach to reduce the data sample volume. The low power analog system is highly useful in this case that adjusts the converter clock rate. The compressed sensing (CS) mechanism, on the other hand, is fundamentally based on packet loss mitigation [9]. This technique is based on the Nyquist sampling technique. Mainly it used to recover the signal with a precise and fine resolution. In order to achieve precision, the CS mechanism relies on the signal of interest of sparse representation of M nonzero elements. Here $M \ll N$ where N is the dimensionality of the signal.

(ii) **Resolution and sampling rate**: Resolution and the sampling rates are the major factors for a signal while they are represented as a digital signal or converted from analog to digital. The resolution is the key factor that allows the number of allowable levels of bit combination through which the data sample can be represented. The rate of sampling is another key factor that ensures the number of samples available at a proper rate so that it can be reconstructed in an easy way. The research has reported that for medical devices like ECG, the sampling frequency of 20 Hz with 2-bit resolution gives optimal performance. Also, this frequency is highly effective in human movement analysis. Some reports show that the 40–50 Hz samples are more effective for human movement analysis. For the ECG graphs in some cases, the 250–500 Hz sampling frequency is also considered.

(iii) Filtering: Filtering technique is highly effective to resolve the noise and distortion related issues. In most of the cases, the medical signals are highly prone to noise. For example, in the case of ECG signal, various noises may affect the signal and may distort the shape of the signal. ECG artifacts are one of the major noises that can be induced in ECG signal results distortion on the actual ECG. This happens due to ambient electrical interference and the electrical signals sent by muscles. The complexity of the filtering highly depends upon the sensors and the application. The filtering like moving average is a very common type of filtering. For personal health monitoring, often the raw accelerometer data gets processed. In such scenarios, the moving average filters perform a removal of the high-frequency artifacts. The major thing about moving average filtering is to choose the window size. In general, the cutoff frequency for moving average filtering is low enough to bypass the unwanted motions. Also, the cutoff frequency should be high enough to grab valid data.

(iv) ECG filtering techniques: There are numerous filtering approaches that have been derived that contribute significantly to the body of the research. Various filtering techniques have been derived to remove the noise signals from the main ECG signal. One of the very basic types of filtering is baseline drifting removal. In this case varying frequency component is very low in quantity. In this case, the abnormal amplitude deflection can be adjusted and removed through this mechanism. The respiratory organ of the body majorly causes such noise which is much greater than the normal waveform. One of the major noise components in ECG is the baseline drift. In this approach, a very low-frequency noise component has been addressed which is primarily produced due to respiratory organs. Types of noise in such cases have a frequency of about 1 Hz. Finite impulse response FIR filtering can also be applied to achieve a 0.5 Hz frequency cutoff with a sampling rate of 250 Hz. Discrete wavelet transformation (DWT) has also been performed at a certain level. This is also considered a filter bank. Median filtering is also an effective methodology to remove DC drift. Moving median filtering is highly used to remove noise from the PQRST component of the wave.

3.4. Biomedical imaging and analysis

Biomedical imaging plays a vital role in clinical applications [10]. The various diseases like abdominal, nerve diseases, muscle disease, and various others give a significant impact on disease identification. A medical image study majorly reveals the location, condition, size, and shape of the organs under concern. Radiology, tomography scans are gaining popularity in various aspects of clinical diagnosis. This technique is of utmost genuine and accurate to understand the behavior of the internal organ and also to identify the affected

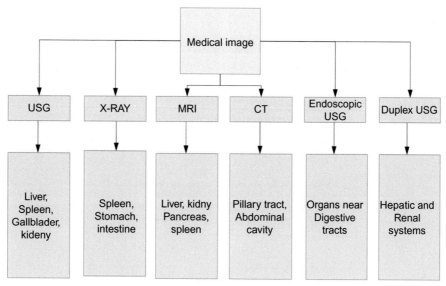

Fig. 3.5 Various medical methodologies for abdominal imaging techniques.

region of the organ, such as lungs, kidney, liver, and many more. Visualization of the major body parts like tissues and ligaments is another major advantage of the medical imaging mechanism. Therefore medical imaging is of utmost use to diagnose diseases and also it supports the understanding of the human body quite critically. Also in order to detect pathological lesions, this technique is highly used. There are various modality imaging techniques that have been considered for abdomen like magnetic resonance image (MRI), computed tomography (CT), positron emission tomography (PET), single-photon emission computed tomography (SPECT), ultrasound, and endoscopy. Fig. 3.5 illustrates various imaging methods that have been incorporated for several medical diagnoses.

3.4.1 Magnetic resonance imaging

MRI scan technique mainly consists of high-intensity magnetic fields and the radio waves which collaboratively perform the imaging [11,12]. A nonionizing and noninvasive system is used to realize the 3D image construction. Some major uses of MRI scans include kidney diseases, liver problems, and abdomen pain. MRI is also very effective to identify and distinguish the difference between tumors and normal tissues. It also assists doctors to understand the size of the tumor, spread, and severity. MRI has both advantages and disadvantages. In many cases, the MRI image is considered better than CT to identify the different tissues and organs. This is because of the varying number of resolutions of the different body parts. There are certain disadvantages of MRI. The main disadvantage involves the metallic part associated with the body. The strong magnetic field created

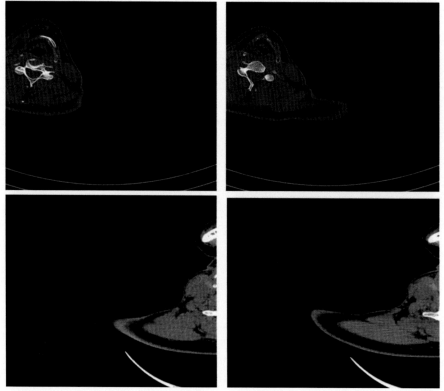

Fig. 3.6 MRI image of the left and the right limbs of the human body.

by the MRI sensing devices may affect implantable devices like pacemakers, in-body sensors, etc. Fig. 3.6 shows the MRI scan image of the left and the right limbs of the human body.

3.4.2 Computed tomography

Computed tomography is sometimes known as CT scan [13,14]. This is basically a X-ray technique through which we can see the cross-sectional images of the precise area of the human body. In this technique, the X-ray device is compactly collimated within a thin slice. This produces a series of 1-dimensional projection of the various angles of the body part. The source and the sensor jointly rotate around the body. As a result, the multiple projections completely result in a 2-dimensional image with significantly reasonable contrast. In order to achieve the final 2-dimensional image, radon transformation has been used. The typical use of the CT scan is to detect an abdominal mass, kidney stone, and abdominal pain. In some cases, abdominal inflammation also required CT scanning. An abdominal CT scan is the relatively safest technique that doesn't have any electromagnetic effects. But as it uses X-ray so may affect human tissues like kidneys and the liver.

3.4.3 Ultrasound-based diagnosis

The ultrasound image is a popular imaging technique that can be done through the sound sensing mechanism [15]. In this case, the sound wave gets reflected in tissues or part of the tissues and returns back to the transducers that capture the signal and convert it to an image based on the frequency and the distance. The system typically operates on a 1–20 MHz frequency range. The high frequency produces high spatial resolution and the ionizing radiation gets reduced. The fundamental clinical application involved intraabdominal imaging of the kidney, liver, stomach, and also flow detection of the blood. Ultrasound is of course a noninvasive and a low-cost detection technique. One of the major drawbacks of this is poor contrast of the soft tissue. So sometimes those tissues are not visible properly in this case. Also, imaging deep inside the body is not possible due to the significant low contrast. Another major imaging technique is the endoscopic ultrasound which uses high-frequency ultrasound waves. Endoscopic ultrasound devices have a thin tube-like sensor probe that can be passed through the mouth or rectum until it reaches the digestive system like stomach, intestine, or large intestine. In this case, sound waves are bounced back and received by the sensor which is further transferred into a computer system. The imaging software in the computer then reconstructs the image from the signal that has been received from the sensor. This technique is very useful to detect cysts, tumors, as well as any diseases of the pancreas, gallbladder, and bile duct. It is also used to get biopsy by applying a thin needle to collect tissue or fluid. This procedure is less harmful because it doesn't consist of radiation so doesn't affect tissues. In some cases, a breathing or bleeding problem is observed in this case. A duplex ultrasound, on the other hand, is another major method that performs recording of sound wave that is reflecting off the blood. Mainly it is used to examine the blood vessels, blood flow in the kidney, abdominal area, liver, and other internal organs.

3.4.4 Abdominal imaging for computer-aided diagnosis

In medical imaging, the computer-aided diagnosis (CAD) is a very popular term nowadays. The main objective of the CAD is to perform medical image analysis and processing that supports the diagnosis. There are several steps involved in medical image analysis such as the transformation of the image. This procedure produces a more precise representation of the image information from the original image which is supposed to be more reliable and precise. Several steps involved before image performing CAD are (a) image formation, (b) image virtualization, (c) image analysis, (d) image management. The formation of the images involves a suitable medium to capture the image in a proper way. In the case of virtualization, the proper image enhancement technique has to be applied. Various artificial intelligence algorithms may be involved in order to enhance the image. Image analysis, on the other hand, mainly emphasizes restoration, segmentation, and classification. Finally, the image management performs the transmission, archival, and retrieval of the image.

3.5. Image enhancement

In order to perform correct diagnosis of medical image, enhancement is of utmost necessity. This procedure corrects the images by reducing the unwanted noise and balance to contrast so that each and every tissue and the parts are clearly visible. There are two major image enhancement mechanisms that are mainly concerned, such as the spatial domain and the frequency domain. One of the popular among them is histogram equalization and bidirectional empirical decomposition. In this case, the overall contrast of the image gets increased significantly. Log transformation-based optimization is also useful to enhance images. This is mainly applicable to CT scan images. This type of image enhancement technique perhaps gives better results in comparison to particle swarm optimization (PSO) based enhancement technique.

Image segmentation is another important issue that has been done after the image enhancement process. During the CT scan or MRI scan, several artifacts have arrived in the image such as partial volume artifacts, streak artifacts. Organs and tissue homogeneity and texture also cause several problems. The gray level distribution model is a stochastic model; an effective technique is to segment the abdominal geometry. The iterative relaxation segmentation has been used that is mainly followed by two distinctive iterative procedures. The fuzzy C-Means segmentation procedure is extremely helpful to perform tumor segmentation.

3.6. Image classification technique

The classification refers to the assignment of the physical object to a predefined category [16]. Mainly this refers to the analysis of the numerical properties and the features of the image captured by various devices. The classification technique is perhaps the final step for the decision making by looking into the category or the class. The class is a predefined object and if the feature of the object falls into a specific class then the medical personnel may make a decision accordingly. Fig. 3.7 shows the classification methodology used in

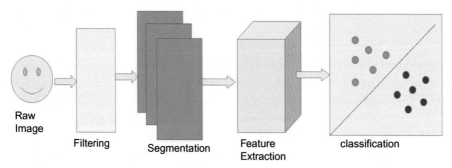

Raw Image   Filtering   Segmentation   Feature Extraction   classification

Fig. 3.7 Image classification procedure by performing the four-layer step approach.

medical images. In the general case, the decision has been made based on binary class or multilevel class. In binary classification, the decision has to be made whether the image belongs to a specific class or not.

Binary classification approaches mainly emphasize on diseases with a probability of positive or negative. Diseases like malignant tumors, cancer, COVID-19, etc. can be classified in this approach. Multilevel classification can also be done to identify the stage of the diseases. The artificial neural network technique is highly used to achieve this.

3.7. Wearable and implantable technologies

One of the most emerging researches on wearable medical sensing technology is the wearable body area network (WBAN). A key component of this mechanism is the BSN or body sensor network. In the recent era, BSN research faces several challenges. The research majorly emphasizes fault tolerance, energy, routing and message transfer, node deployment, data aggregation, feature extraction, and data classification. There are numerous sensors that can be applied in the various parts of the body so that they can monitor the specified regions. The various sensors that are illustrated in Fig. 3.8 have been described as follows.

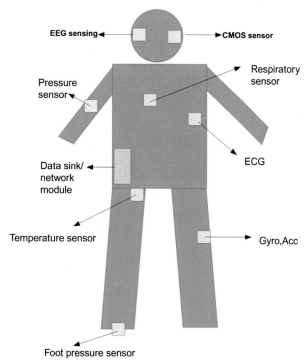

Fig. 3.8 Various sensors and their deployment position.

(a) Accelerometer: It primarily measures the acceleration of the body component. While walking or moving the arm and legs the value of the accelerometer may change and the sensor data get generated. This sensor has a vital use in the detection of energy consumption in the human body. The sensor also gives an idea of the activity of the body, motion pattern, and intensity.

(b) Pressure sensor: In the case of the BSN network, the pressure sensor is mainly used to perform blood pressure measurement and monitor the ups and downs of the pressure. It also gathers the data and performs the store and forward of the data through the network channel. In general, a pressure sensor is difficult to integrate with other modules. The energy expenditure can be measured using the collected data by performing feature extraction and classification.

(c) Respiratory sensor: This is a combination of all sensors like gyro, pressure, accelerometer, etc. This works based on the data gathered by the devices like chest and abdomen expansion and contraction. There are various work has been done on breathing feedback based wearable devices and sensor systems to aggregate and fuse the respiratory data.

3.8. Conclusion

The chapter mainly shows various aspects of sensing and related issues along with signal processing, imaging, and artificial intelligence techniques applied to the sensor data. Various node deployment methodologies have also been addressed and various aspects of applications of imaging, signal processing, and intelligent techniques are mainly emphasized.

References

[1] S. Laxman, P.S. Sastry, A survey of temporal data mining, Sadhana 31 (2) (2006) 173–198.

[2] W.W. Wei, Time series analysis, in: The Oxford Handbook of Quantitative Methods in Psychology: Vol. 2, 2006.

[3] R.E. Challis, Biomedical signal processing (in four parts). Part 2. The frequency transforms and their inter-relationships, Med. Biol. Eng. Comput. 29 (1) (1991) 1–17.

[4] R.E. Kass, V. Ventura, C. Cai, Statistical smoothing of neuronal data, Network-Comput. Neural Syst. 14 (1) (2003) 5–16.

[5] H. Akaike, Fitting autoregressive models for prediction, Ann. Inst. Stat. Math. 21 (1) (1969) 243–247.

[6] D. Piccolo, A distance measure for classifying ARIMA models, J. Time Ser. Anal. 11 (2) (1990) 153–164.

[7] M. Muzammal, R. Talat, A.H. Sodhro, S. Pirbhulal, A multi-sensor data fusion enabled ensemble approach for medical data from body sensor networks, Inf. Fusion 53 (2020) 155–164.

[8] M. Allsworth, Owlstone Medical Ltd, Improved breath sampling device and method, 2020. U.S. Patent Application 16/493,366.

[9] A.Q. Wang, A.V. Dalca, M.R. Sabuncu, Neural network-based reconstruction in compressed sensing MRI without fully-sampled training data, in: International Workshop on Machine Learning for Medical Image Reconstruction, Springer, Cham, 2020, pp. 27–37.

[10] S.J. Fong, G. Li, N. Dey, R.G. Crespo, E. Herrera-Viedma, Composite Monte Carlo decision making under high uncertainty of novel coronavirus epidemic using hybridized deep learning and fuzzy rule induction, Appl. Soft Comput. 93 (2020) 106282.

[11] A.G. Christodoulou, S.G. Lingala, Accelerated dynamic magnetic resonance imaging using learned representations: a new frontier in biomedical imaging, IEEE Signal Process. Mag. 37 (1) (2020) 83–93.

[12] L.L. Wald, P.C. McDaniel, T. Witzel, J.P. Stockmann, C.Z. Cooley, Low-cost and portable MRI, J. Magn. Reson. Imaging 52 (3) (2020) 686–696.

[13] V. Rajinikanth, N. Dey, A.N.J. Raj, A.E. Hassanien, K.C. Santosh, N. Raja, Harmony-search and otsu based system for coronavirus disease (COVID-19) detection using lung CT scan images, arXiv preprint arXiv:2004.03431 (2020).

[14] N. Dey, J. Chaki, L. Moraru, S. Fong, X.-S. Yang, Firefly algorithm and its variants in digital image processing: a comprehensive review, in: Applications of Firefly Algorithm and its Variants, Springer, Singapore, 2020, pp. 1–28.

[15] D.T. Nguyen, J.K. Kang, T.D. Pham, G. Batchuluun, K.R. Park, Ultrasound image-based diagnosis of malignant thyroid nodule using artificial intelligence, Sensors 20 (7) (2020) 1822.

[16] Z. Huang, X. Zhu, M. Ding, X. Zhang, Medical image classification using a light-weighted hybrid neural network based on PCANet and DenseNet, IEEE Access 8 (2020) 24697–24712.

CHAPTER 4

Sensor data analysis

Contents

4.1. Machine learning preliminaries

Machine learning and artificial intelligence are the two important topics in the modern industrial revolution. In a naïve way, we can simply define machine learning as the task of categorizing similar data into groups and drawing inferences from it. From the drawn inferences, the task lies in categorizing or identifying the new set of inputs and to which group or label they will belong. Artificial intelligence is a much bigger and broader concept, which actually comprises applying machine learning procedures in a much smarter way.

In order to define a machine learning model, a set of learning procedure is given, along with the data to the computer to learn and draw inferences. The machine learning procedure is made to adjust with respect to data provided for learning. The term that we usually use is "**fit the model on the data**" or that "**model is trained using the data.**" The data can be continuous or data can be labeled or categorical. A simple overview of the type of data on which machine learning is performed can be demonstrated with the following example. Let us consider the ECG data; it comprises measures of electrical signals or activity of the change of heart over time. Hence, the data generated is continuous.

Biomedical Sensors and Smart Sensing
https://doi.org/10.1016/B978-0-12-822856-2.00002-2

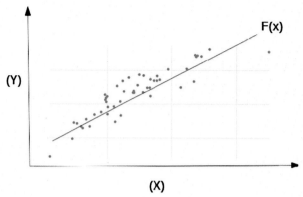

Fig. 4.1 Line fitting on a scatter plot.

Continuous data [1] falls under the class of regression problem. Regression [2] procedure deals with drawing relationships between the dependent and the independent variables. Given an unknown X, we have to predict Y. Let us consider a simple linear regression problem (Fig. 4.1), where X is the only feature or attribute on which we have to estimate the Y value. A depiction of a simple line is a curve fitted across a scatter plot of the data. Now the function can have a different form; hence, the nature of the fitted curve can be of any shape, which forms the basis of the learning mechanism.

A simple overview of linear regression is given by the equation

$$Y = \beta_0 + \beta_1 X + \varepsilon \qquad (4.1)$$

$$Y = \beta_0 + \beta_1 X^1 + + \beta_1 X^2 + \ldots + \beta_p X^p + \varepsilon \qquad (4.1.1)$$

The expected value of Y given X:

$$E(Y|X) = \beta_0 + \beta_1 X + \varepsilon \qquad (4.1.2)$$

$$E(Y|X) = \beta_0 + \beta_1 X \qquad (4.1.3)$$

We have to find the β_0' and β_1' for simple linear regression. For multiple linear regression, we have n variables, *an* independent variables $\{x_1, x_2, \ldots x_n\}$. So we have "n" predictor variables. $E(Y|X)$ is the expected value Y given X, $E(Y|X)$ follows the equation we are trying to estimate $\beta_0', \beta_1', \ldots, \beta_n'$. Regression can be classified into simple and multiple as depicted in Fig. 4.2. In order to perform linear regression with multiple attributes, the following is the equation:

$$E(Y|X) = \beta_0' + \beta_1' X^1 + + \beta_1' X^2 + \ldots + \beta_p' X^p \qquad (4.2)$$

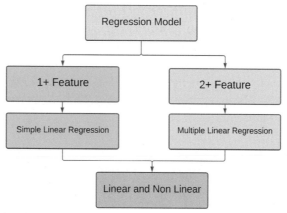

Fig. 4.2 Regression classification.

For both linear and multiple linear regression, the main motive lies in reducing the sum of squared errors, which is achieved through various algorithms. The assumption about errors ε is denoted as follows for simple linear regression:

Datapoints:

$$
\begin{aligned}
d_1: \quad & Y_1 = \beta_0 + \beta_1 X_1 + \varepsilon_1 & (4.3) \\
d_2: \quad & Y_2 = \beta_0 + \beta_1 X_2 + \varepsilon_2 & (4.3.1) \\
d_3: \quad & Y_1 = \beta_0 + \beta_1 X_3 + \varepsilon_3 & (4.3.2) \\
& \dots\dots\dots\dots \\
d_n: \quad & Y_1 = \beta_0 + \beta_1 X_n + \varepsilon_n & (4.3.3)
\end{aligned}
$$

$$\varepsilon = \left(\text{Actual value} - \text{calculated value}\right)$$

Hence,

$$\varepsilon = \left(Y_i - \left(\beta_0 + \beta_1 X_i\right)\right) \tag{4.4}$$

The sum of squared errors is estimated to consider the sum of squared differences within each observation and consider the mean of the data sample. The main objective is to consider the variance between the clusters of data. If the cluster of data is identical, then the sum of squared errors is 0; thus $E(\text{mean}(\varepsilon))$. The sum of squared errors (SSE) is represented as follows:

$$SSE = \sum_{i=0}^{n} \left(Y_i - \left(\beta_0 + \beta_1 X_i\right)\right)^2 \tag{4.5}$$

The objective of regression lies in drawing the curve, which best fits the points. The curve is drawn in order to reduce the sum of the squared error to the minimum value. Hence, finding the value for β is important. Regression has been extensively used and studied and applied in various domains of engineering. It is widely used as a statistical analysis tool in biomedical research. One such work can be found in [3] where the researchers have applied regression in estimating the optical properties of biological material, i.e., in the field of biomedical optics. The authors have used polynomial regression for providing solutions to the problem. Another important work has been discussed in [4] where the authors have claimed to have performed segmentation of ECG signals by use of linear regression procedure. They have used linear regression to get the maximum and minimum from an acquired ECG wave.

Pattern classification [5], on the other hand, can be divided into statistical and structural. Each approach employs different techniques to implement the description and classification tasks. There are numerous hybrid approaches that combine statistical and structural approaches in a pattern classification problem. Pattern classification is taken from various accepted concepts in decision theory to separate data into various groups based on the quantitative features of the data. There are a wide variety of statistical techniques that are rigorously used in the extracting the features. Some of the widely used statistical feature extraction techniques include standard deviation and mean estimation, frequency calculation, various transformations, etc. The features that are extracted from every object for the very task of classification are grouped into arrays called the feature arrays. Every feature describes a particular characteristic of the data. The input to the classification task is the features and their respective data. Various statistical approaches are also used for the task of classification, which majorly focus on similarity of the data (e.g., k-NN [6], template matching, etc.), probability (e.g., Naïve Bayes [7]), boundaries (e.g., decision trees [8], neural networks [9], deep learning), and clustering-based (e.g., k-means [10], hierarchical [11], DBSCAN [12], etc.).

In the present biomedical research domain, machine learning plays a key role in shaping the way data is processed. Although before the actual data is analyzed, relevant features have to be extracted from the raw data depending on the application-specific need. A simple overview of the use of machine learning for mood modeling classification in psychiatric research is done by collecting recordings of the patient with the use of a smartphone. Hence, the domain of biomedical research is not limited to the creation of a dedicated system for analysis. Any everyday device can be modeled to work for the generic cause in the domain of healthcare. The majority of the data that are processed are ECG, EEG, MRI images, X-ray, ultrasound, etc.

One of the major problems when evaluating a model is that it is very difficult to know how much error is induced due to noise versus the insufficiency of the model. The objective of the machine learning model should be that it should give better results than any human-based analysis. If a human-generated model generates results and the results from the machine learning model are different, then it may be because the function which is modeled is misguided.

4.2. Feature engineering

In the present scenario, we must have come across the term "**Data is oil**," the term is correct as well as incorrect. Data is generated across various sensors in and around the world. The generated data are in the raw format, which has to be parsed and preprocessed before the machine can actually learn from it. Further, it is important to select the best variables that would help to construct the mapping between the input and the output. The variables are called *"features,"* and it is crucial to select the best set of features that best defines the model. Feature engineering is the most important task, and researchers have invested a lot of time in coming up with the best solutions for the problem. We must have come across the term *"Feature Selection"* as well as *"Feature Extraction."* Both the terms seem similar but there is a slight difference. Feature selection involves the task of filtering out redundant and unimportant features from the set of feature pool, hence keeping or selecting a subset of features from the old feature set. Feature extraction, on the other hand, involves transforming and modifying, and hence the word extracts a new set of features, thereby creating a new set of attributes to carry out the learning process.

Feature selection [13] is broadly classified into a supervised selection and unsupervised selection. A basic overview of the supervised feature selection procedure is correlation, variance, mean, etc. The focus of any feature selection procedure is to select the best subsets of features among the feature set removing irrelevant and redundant features in the process.

$$\text{Correlation}(i) = \frac{Cov(x_i, Y)}{\sqrt{\text{var}(x_i) * \text{var}(Y)}} \qquad (4.6)$$

$$\text{var}(x_i) = \frac{\sum (x_i - x')^2}{n} \qquad (4.7)$$

Features can be modeled in a way where the correlation is estimated for each of the columns and is ranked in a manner according to application-specific need. Though, in a classification process, it is required to have features with high variance and low correlation. This directly impacts the classification accuracy and the error metric of any classification process. Numerous other selection criteria exists such as information gain where the features are ranked on the basis of the amount of information collected.

Unsupervised feature selection doesn't require the need of labeling the data. Unsupervised learning finds patterns within the data without human supervision. Principal component analysis (PCA) [14,15] is a widely used approach for dimensionality reduction. PCA is a statistical approach that selects the features by taking the axes of the great variance of the data. PCA performs rotation of data from one coordinate system to another. One of the common mistakes is the application of PCA on continuous data. Though, it is possible to use PCA on discrete or categorical data. It is not recommended to use PCA if the variables don't belong to a coordinate plane. An application of the PCA approach with classification done using Random Forest classifier is shown below (part of the sklearn library of Python). Here, only one component is considered for classification and the output is showing an accuracy of 85.34%.

Code: *PCA in Python*

```
importnumpy as np
import pandas as pd
fromsklearn.model_selection import train_test_split
fromsklearn.decomposition import PCA
fromsklearn.preprocessing import StandardScaler
fromsklearn.decomposition import PCA
fromsklearn.ensemble import RandomForestClassifier
fromsklearn.metrics import confusion_matrix
fromsklearn.metrics import accuracy_score
fromsklearn.preprocessing import StandardScaler


names = ['x1', 'x2', 'x3', 'x4',… …, 'Class']
dataset = pd.read_csv(data.csv)

X = dataset.drop('Class', 1)
y = dataset['Class']


# Splitting the dataset into the Training set and Test set


X_train,   X_test,   y_train,   y_test   =   train_test_split(X,   y,   test_size=0.3,
random_state=0)


standScal = StandardScaler()
X_train = standScal.fit_transform(X_train)
X_test = standScal.transform(X_test)

# Principal component analysis
pcomp1 = PCA(n_components=1)
X_train = pcomp1.fit_transform(X_train)
X_test = pcomp1.transform(X_test)


Class_C = RandomForestClassifier(max_depth=k, random_state=0)
Class_C.fit(X_train, y_train)

# Result Prediction of classifier on one component
pred_Y = Class_C.predict(X_test)

#Classifier Accuracy Printing
print('Classifier Accuracy: ', accuracy_score(y_test, pred_Y))

Output:

Classifier Accuracy:85.34
```

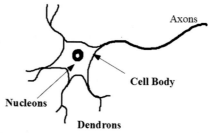

Fig. 4.3 Nerve cell/neuron diagram.

4.3. Perceptron learning

The human brain contains billions of neuron nodes interconnected with each other processing information and making predictions. A simple diagram depicting the neuron is given in Fig. 4.3. The neuron comprises nucleons, dendrons, cell body, and axons. Nucleons are the nucleus of the nerve cell. The dendrites are responsible for propagating the electrochemical stimulation received from other nerve cells. Artificial neural network (ANN) [16], as we know of today, is one of the most important areas in the domain of machine learning, which is performed by modeling the human brain and simulating the way in which the brain analyzes data. The ANN procedure incorporates two of the fundamental components of a biological neural network.

(a) Nodes—Neurons
(b) Weights—Synapses

The basic unit of the neural network is called a "**Perceptron.**" The Perceptron [17] is a more generalized computational modeling of a neuron. It takes in inputs and assigns weights to the inputs and fires some results based on some threshold parameter.

The diagrammatic representation of a perceptron is shown in Fig. 4.4. Now, $\phi(z)$ can be a thresholding function that triggers binary output depending on the value. If

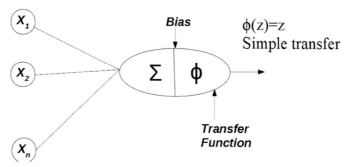

Fig. 4.4 Perceptron diagram.

the Σ is greater than the threshold, then it might output 1; else, it outputs 0. $\phi(z)$ can also be a sigmoid function that looks like an *S-shaped* curve and is a real bounded differentiable function or a rectified linear unit (ReLU) [18]. A single perceptron can be used to develop a linearly separable function that considers both Boolean value and real-value inputs and assigns weights with some threshold or bias. A simple perceptron can be modeled as follows. A code depicting a single neuron equation is given below.

$$y = \sum_{i=1}^{n} w_i x_i + b \qquad (4.8)$$

Code: *Single perceptron*
```
input = [2,2,3,2.5]
weight1 = [0.2,0.4,-0.5,1.0]
weight2 = [0.23,-0.91,0.26,-0.5]
weight3 = [-0.26,-0.27,0.27,0.87]
bias1 = 2
bias2 = 3
bias3 = 0.5
output = [input[0]*weight1[0]+input[1]*weight1[1]+input[2]
*weight1[2]+input[3]*weight1[3]+bias1,
    input[0]*weight2[0]+input[1]*weight2[1]+input[2]
*weight2[2]+input[3]*weight2[3]+bias2,
input[0]*weight3[0]+input[1]*weight3[1]+input[2]
*weight3[2]+input[3]*weight3[3]+bias3]
print(output)
```
Output
```
[4.2, 1.17, 2.425]
```
The limitation of the perceptron is that it has monotonicity properties. If a link has positive activation, it can only increase as the input value increases. It cannot represent functions where input interaction can cancel each other. So each input is individual with the neuron so it cannot handle the interaction between neurons. Hence, the interconnection of more than one perceptron, i.e., a multilayer perceptron or what we called a neural network, is suitable for the nonlinear function. Multilayer structures can express interaction among the inputs. The first layer of the multilayer structure is the input layer, the second layer is the hidden unit, and the last layer is the output layer. When we say that a 2-layer neural network, it is actually 2 hidden layers. The input and the output layer are always present. The hidden units are the key perceptrons used for the learning process. The estimation or prediction is done by the forward

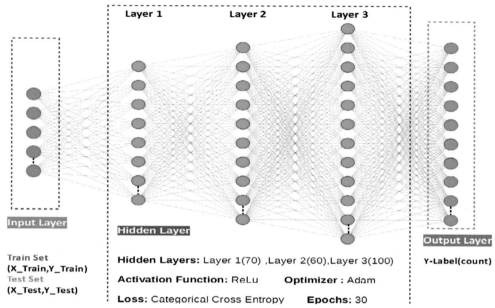

Fig. 4.5 Three-layer neural network representation.

propagation. In the forward propagation through the network, the input data is forward propagated; i.e., the summation with the weights and biases are estimated in each neuron, and with the selected activation function, the predictions are made. With the backpropagation, the prediction estimation is done to get an understanding of the deviation from the actual class label. The loss estimated is backward propagated to correct the weights involved in the neural net processing. Some of the popularly used loss functions are the cross–entropy or log loss, mean square error, etc. In Fig. 4.5, a depiction of a 3-layer neural network is done. The first layer as we can see contains 70 nodes, the second contains 60 nodes, and the last hidden layer contains 100 nodes. The activation function used in feed-forward prediction is rectified linear unit or ReLU. The loss function used is categorical cross-entropy, as the labels are categorical data. Given below is a code showing the implementation of the 3-layer neural network with the given setting in Fig. 4.5. Our code is created using the TensorFlow and Keras library of Python. The output layer has softmax activation that predicts multinomial probability; specifically, it converts weighted sum values into probability values. With respect to each label, a probability is generated.

Code: *Feed-forward ANN with activation function*

```
importtensorflow as tf
fromtensorflow.keras.models import Sequential
fromtensorflow.keras.layers import Dense
fromtensorflow.keras.losses import categorical_crossentropy
fromtensorflow.keras.utils import to_categorical
import pandas as pd
importnumpy as np



df = pd.read_csv("Dataset.csv")

X = df.drop(columns='Label')
y = np.array(df.loc[:,'Label'].tolist())

#Dividing the dataset into 70% training and 30% testing or validation set.
train_X, test_X, train_y, test_y = train_test_split(X, y, train_size=0.7)



defbuild_network(features, output, *args):
model = Sequential()

    #Input Layer Creation
model.add(Dense(args[0], input_shape=(features,), activation='relu'))

#Three Hidden Layers 70,60,100
for a in args[1:]:
model.add(Dense(a, activation='relu'))

#Output Layer Creation
model.add(Dense(output, activation='softmax'))

model.compile(optimizer='adam',loss='categorical_crossentropy',
metrics=['accuracy'])

return model



#Building the model layers by calling the function build_network
# n is the number of output labels
full_model = build_network(len(df.columns[1:]), n, 70, 60, 100)

#Training the model
temp          =          full_model.fit(train_X,train_y,epochs=30,validation_split=0.2,
batch_size=12)


P = full_model.evaluate(x=test_X, y=test_y, batch_size=12)
```

4.4. Application of machine learning on ECG data

4.4.1 ECG raw data classification

As we know, ECG is the measurement of the electrical property of the heartbeats and is one of the most important diagnosing tools for various heart diseases. It is very essential in modern-day healthcare to predict the diseases before the occurrence of the actual disease or to predict diseases that are not visible to the naked eye. It is very important to extract and select features in order to perform analysis. Many researchers have detected QRS complexes [19] with help of signal characteristics like high frequency and amplitudes, etc. Principal component analysis, particle swarm optimization [20], and many other supervised approaches are widely used for attribute extraction and classification. However, it still remains a challenge in extracting the exact features. ECG signal has different components of different frequencies. One of the major issues in classifying ECG data is the lack of standardization of features in ECG, i.e., the attributes with which the classification task has to be carried out. Due to the variability of ECG attributes and variability in ECG waveforms of patients, there is no fixed or optimal classification rule. One of the major issues in developing the most appropriate classifier that is capable of predicting arrhythmia in a real-time manner is the variation of beats, which varies highly from person to person. Hence, the training algorithm has to be modeled accordingly. ECG signal classification is very essential and the most important applications of classifying ECG signals are in detecting abnormality type and diagnosing a patient by performing computational analysis.

The steps that are involved in the classification procedure are:
1. Preprocessing and feature extraction
2. Normalization
3. Classification

Preprocessing involves various signal-level modifications, filtration, and transformation. It is mainly used to remove noise from the signals. For the task of preprocessing low-pass filters, linear phase high-pass filters, etc. are widely used. In order to adjust signal baseline, filters like a median filter, high-pass linear phase filter, etc. are used. During the feature extraction phase, it is very important to detect the R-points in the ECG signal. The R-peaks in the signal are detected using wavelet transform (WT). Discrete wavelet transform [21] is one of the most widely used approaches in the feature extraction procedure. Some of the other approaches like continuous wavelet transform (CWT) [22], discrete cosine transform (DCT), and discrete Fourier transform (DFT) [23] are also widely used. For the normalization process, scaling techniques such as Z-score [24] and unity standard deviation (SD) are used.

Notable work on heart rate classification is depicted in the paper [25]. The work considers heart rate variability for the prediction of various diseases. Within the successive R-peaks, the time interval is considered with which the heart rate is evaluated from the ECG signal. This is also called the R-R interval. The heart rate at which it is sampled is 200 samples/s, which is plotted against a time scale (Fig. 4.6A and B). The prediction is performed using an artificial neural network. The neural network (Fig. 4.7A), as we know, is created using the interconnection between the neurons. The power of a neural network comes from the number of the interconnection and the number of layers that are

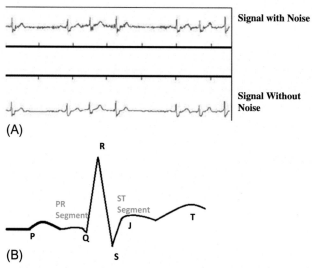

Fig. 4.6 (A) ECG data with noise and without noise. (B) PR and ST segment.

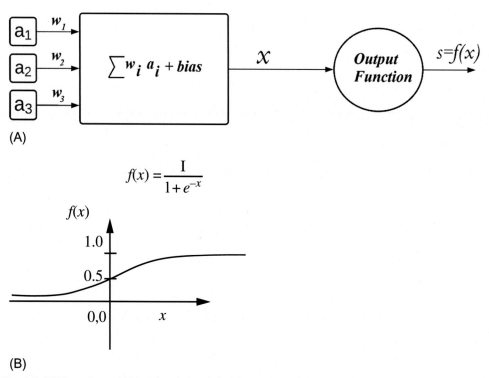

$$f(x) = \frac{1}{1+e^{-x}}$$

Fig. 4.7 (A) Three input ANN with weights defined as w1, w2, and w3. (B) Sigmoid activation function.

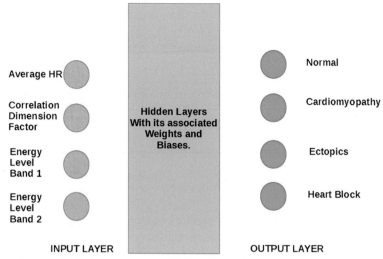

Fig 4.8 Neural network block diagram opted for ECG classification.

the hidden layers. Choosing the correct neuron activation function plays a very important part in the design of the ANN. Usually, a sigmoid activation function (Fig. 4.7B) is used for the feed-forward neural network. The backpropagation helps us to correct the errors.

The input layer of the neural network is the feature values. The hidden layers process the data by assigning weights with the input data and passing through an activation function to direct it toward an output label. Let us consider the output to have four labels; hence, there are four nodes in the output layer. According to the data in the features, they are labeled into any one of the four labels. The labels are *normal, cardiomyopathy, ectopic, and heart block,* which are depicted in Fig. 4.8. They have considered four attributes or features, which are average heart rate, energy content in the frequency band (33.3–100 Hz), energy content in the frequency band (66.7–100 Hz), and correlation dimension factor.

4.4.2 ECG image classification

An approach utilizing the ECG image data is given by Jun et al. [25]. The authors have used the MIT-BIH Arrhythmia database (*https://www.physionet.org/content/mitdb/1.0.0/*). The ECG images were preprocessed and converted into 128×128 grayscale images. The objective of the approach is to predict heartbeats, and eight different types of heartbeats were categorized and used for prediction. A diagrammatic representation of the steps involved is given in Fig. 4.9.

The objective of a convolutional neural network (CNN) is to take in pixels of the images as the input and form the learn mechanism by the very same process a normal

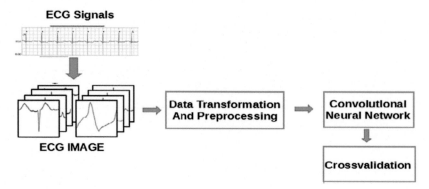

Fig. 4.9 ECG signal classification from ECG images trained with a CNN.

neural network does by modifying the weights and biases. Now the images contain a large number of pixels, and the duty of the CNN is to reduce the number of pixels through the use of various filters or popularly called the kernels. A demonstration of the block diagram of a CNN is given in Fig. 4.10. The first layer of the CNN is called the convolution layer. The convolution layer is responsible for connecting the local area of the input neuron instead of a fully connected network. This immensely reduces the number of attributes for processing. After the convolution layer, we have the pooling layer. The pooling layer is the part where the kernels are applied for reducing the dimension of the pixels. Two of the most popular pooling approaches are max pooling and

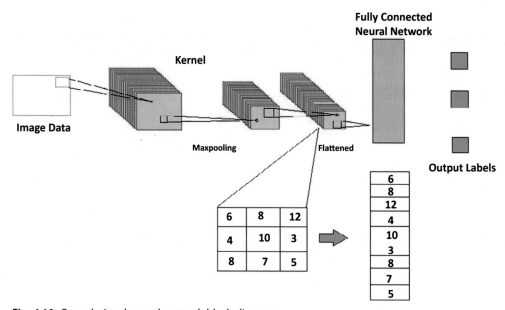

Fig. 4.10 Convolutional neural network block diagram.

average pooling. In max pooling, the highest value in the kernel scope is considered, while in the case of average pooling, the average value in the portion of the image is evaluated. At the very least, the whole of the output from the previous layers is flattened, which is actually a 1D form of the 2D vector and is passed through a regular neural network for the prediction process.

4.4.3 ECG sound segmentation

The stethoscope is a widely used device to hear and understand the heartbeat pattern and detect various abnormalities in the heart and in the chest. Various conditions like murmurs and aberrations of heart sound can be collected, and analysis can be done. Phonocardiogram or PCG in short is a device used to collect sound data from the heart. Usually, the PCG signals consist of two heart sound clubbed together, namely, *lup* and *dup*. Various diseases or abnormalities can cause some extra sound or signal in between these two signals. These perturbation signals are mainly used to identify various heart diseases. Segmentation of these PCG signals is very important. In order to extract or segment the signal, clustering methods can be used. A very notable work using ECG sound is given in [26]. The authors have utilized K-means clustering to indicate a single detected cycle of ECG. For the detection of peaks or R-peaks as known from the ECG plot, homomorphic filtering can be used. A procedure called peak conditioning is used to remove the peaks that did not contain the *lup* and *dup*. K-means clustering of the time intervals between peak values is utilized to indicate the occurrence of a very single cardiac cycle and also to consider the missed cycle. The feature extraction is performed by the use of wavelet transform to obtain various features from PCG signals. The classification approach used is to grow and learn neural networks whose working is detailed in [26].

4.5. Ambient medical data processing

The objective of the present-day need is the presence of ubiquitous processing in our surroundings. Ambient intelligence is a new technological advancement, in which people are empowered through the digital era that is aware of their presence and context and is sensitive, adaptive, and responsive to their needs. In the domain of medical data processing, the sensors form the basis of the data gathering process. The systems are constructed with the objective of inferring knowledge from the data for decision making. The characteristic of an ambient technology can be listed as:

(1) Scalable and adaptive;
(2) Local and remote processing;
(3) Ubiquitous nature;
(4) Personalization;
(5) Context awareness.

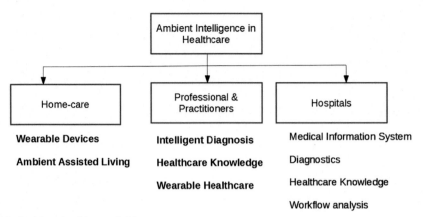

Fig. 4.11 Ambient healthcare fields.

Ambient technology [27] in medical healthcare is built with the objective to improve the real-time monitoring of individuals and collect and process the information locally as well as remotely. The fields where ambient intelligence can be broadly used are home care, hospitals, and medical practitioners (Fig. 4.11). Ambient technologies have greatly affected the way the patients are monitored in hospitals paving the way toward smart hospital rooms, which provide great support to the patients as well as the medical support team. In order to timely perform medical intervention for healthy individuals or living, ambient intelligence focuses on incorporating intelligent processing into the healthcare environment. An ambient system contains various sensors for the recognition of shapes, various activities, sound, etc. In order to increase the efficiency, the data collected from the various sensors have to be preprocessed and simplified. The traditional computing models are no longer needed in the modern computing paradigm. Sensors and micro-controllers are integrated into almost every object that works together for supporting the individuals. Machine learning and artificial intelligence are widely used for performing analysis that allows successful understanding and diagnosis of various symptoms. Such information is gathered together and adapted with the environment users need.

We have discussed the key framework of any biomedical system, i.e., the sensors; another framework supporting any ambient system is the wired and wireless networks. Technologies such as Wi-Fi, Bluetooth, RF, and Zigbee are widely used, which are incorporated for the transmission and reception of data from the sensor devices. The low–data rate IEEE 802.15.4 technology (ZigBee) is very popular in medical monitoring systems for its low transmitter power. Though, the systems that are built using the Zigbee platform are perturbed by strong interference from the WLANs that are ubiquitously present as Wi-Fi. In a hospital administration, it becomes really challenging as there involves a lot of other systems that might interfere with the system. Recently, low–data rate, ultra-wideband (UWB) technology is another popular approach that could be used for body-area network applications because of its controlled and regulated low transmitter power.

Sensor network forms the backbone of data transfer operation from the sensors to the processing device. The wireless sensor network (WSN) transmits the data gathered from the sensors from one node to other nodes wirelessly. WSN is a collection of assorted sensor nodes, and few design considerations are considered, which include the size of the sensor nodes, consumption of power, etc. The data relayed by the sensor nodes are processed in a server. The processing is also distributed into three design considerations, namely, edge-level, fog-level, and cloud-level processing, which is discussed in the latter part of this book. The gathered information is often used to construct *Medical Information Systems*. Medical Information Systems are secure digital repositories where various tests, medical images, gathered signal data like ECG, EEG, etc. are stored as an electronic health record (EHR). The records are shared among all the authorized personals in the hospitals for analysis. The present ambient intelligence in medical healthcare objective lies in sharing the EHR record, which can be made accessible from anywhere and at any time.

4.5.1 Wearable healthcare

In the present era, wearable technologies like smartwatches, smartbands, smartshoes, and smartglasses have divided into the market and people are adapting to the new technological paradigm of real-time monitoring. A simple smartband or watch is able to detect the number of steps monitor vitals through photoplethysmography [28]. A technology to optically obtain plethysmogram is majorly used to check blood volume change and hence extract information just as the heartbeat. The pulse oximeter is another device that uses photoplethysmography for heart rate analysis. As wearable technologies provide continuous monitoring of human physical activities, chemical change, and behavioral parameters, it has found its exclusive usage in the domain of healthcare analytics providing innovative solutions to healthcare problems. Wearable devices used for patient management and disease diagnosis directly impact clinical decision making. The most common data that are collected are the temperature of the body, gait changes, electrocardiogram, photoplethysmogram, sweat, etc. The data gathered from the sensors act as the oil in analytics for healthcare. Machine learning augmented with wearable technologies is used for oncology, radiology, surgery, rheumatology, neurology, cardiology, etc. domains. As the up-gradation of technology takes place, the analytics outcome and the applications are bringing an enormous impact in transforming and the industry.

4.5.2 Wireless body area network

The sensors used in gathering data agglomerate the data and transfer physical, chemical, or biological information to control or processing units placed within one's body or to a remote base station. WBAN can be used as an implantable sensor node or in a wearable form to gather data. The development of a WBAN [29] involves tackling the classical issues pertaining to wireless sensor networks like routing, topology selection, placement problem, etc. Another major issue is energy consumption. In any BAN framework, the emitted power must be controlled in order to avoid damage to the tissues due to over-heating, and also to preserve the battery power. Data transfer in wireless sensor networks can be categorized as unicast, multicast, broadcast, and convergecast. The unicast mechanism works in the manner of one-to-one communication or delivery of data. A simple transfer of data from one sensor node to another or to a base station is a unicast data transfer. The MAC protocols support the time division multiplex mechanism of data transfer with very basic and lightweight versions of operation that are supported by the sensor nodes. Multicast is usually used to send the data to a group of base stations or to multiple sensor nodes. Emphasis is given to various metrics like bandwidth, data packet latency, etc. Broadcast data transfer is carried out by transferring the data to all the nodes in the network. A chain-based transfer is performed due to which all the nodes and even the intermediate and the final base station receive the data. Convergecasting, on the other hand, deals with the gathering of sensor data toward a single sink node.

In the present healthcare scenario, the sensors that gather the information pertaining to one individual are not fully wearable. Some of them have heavy wiring (Fig. 4.12A) and are bulkier in design. Thus in WBAN, it is very essential to miniaturize the sensor nodes and they should perform in low power. The power consumption depends majorly on the transmission methodology opted. It is desirable that the WBAN platform should be built on a wireless platform (Fig. 4.12B); this ensures low power consumption and reducing the transmitted power.

WBAN can be implemented to connect as strap in the hand or leg containing sensors for vitals monitoring. Sleeping disorder throughout the world affects the productivity and health of an individual throughout the world. Monitoring the sleep pattern is very vital in recent times to help the medical practitioners better aid the patients. The test *polysomnography* is used where various recordings of biopotentials are done overnight using sensors and WBAN based systems are developed which reduces the use of wires and instead

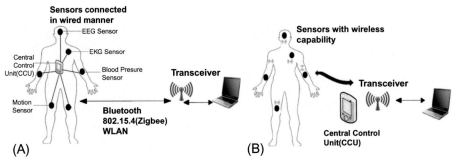

Fig. 4.12 (A) WBAN connected to the central control unit (CCU) in a wired manner. (B) WBAN connection with CCU in a wireless manner.

transmits the data over sensor networks. WBANs are also used for gait analysis, patient monitoring in gluecocellphone technology for monitoring diabetic patients.

4.5.2.1 WBAN implementation for patient monitoring

Any WBAN implementation should have some design perspective in order to make it robust for its usage in medical environments. The communication or transfer of data in any WBAN should happen wirelessly. The sensors used in the implementation should be low cost and less size. The antenna used for communication should be a small high gain antenna to increase the reliability of transmission. An important issue as discussed before is that WBAN gets perturbed from other nearby wireless signals; hence, each sensor signal should work with a different frequency and should be optimized accordingly. Gateways are constructed for remote monitoring and analysis over a very long range. Implementation of data transfer and communication framework is shown in Fig. 4.13A and B.

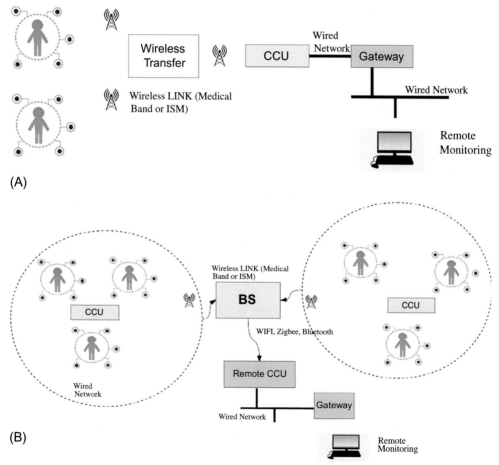

Fig. 4.13 (A) Wireless remote monitoring where sensors are equipped with wireless transceivers. (B) Convergecasting approach from multiple CCU towards a single base state station (BS).

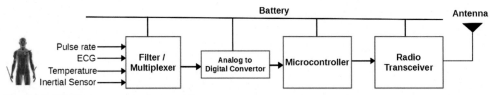

Fig. 4.14 WBAN data processing workflow.

For any generalized development of WBAN and performing real-time monitoring, dedicated sensor nodes with their circuits and interconnection have to be developed. In Fig. 4.14, we can observe a block diagram of the framework for a WBAN sensor interconnection. We can observe that the collected sensor data pertaining to pulse, ECG, temperature, and inertia are passed to the multiplexer and filter circuit. The raw data are passed to an analog-to-digital converter, which is stored and processed in a microcontroller device. PIC16F785 is an 8-bit microcontroller unit (MCU), AVR is another variant popularly used for the implementation of WBAN. The radio transceiver helps in transmitting and receiving packets wirelessly.

4.6. Conclusion

This chapter gives a brief introduction into machine learning and neural networks. This chapter systematically gives an overview of the learning models and discusses some of the work used in healthcare domain for the prediction of various diseases and ailments. The importance of ambient healthcare and the overview of WBAN is also discussed in this chapter. It can be observed that ambient assisted technologies in medical healthcare has greatly impacted the real-time monitoring of people. And the collections and remote analysis has paved way toward the building of training datasets that are used for various diagnostic purposes.

References

[1] W.J. Krzanowski, D.J. Hand, ROC Curves for Continuous Data, CRC Press, 2009.
[2] S. Chatterjee, A.S. Hadi, Regression Analysis by Example, John Wiley & Sons, 2015.
[3] J.S. Dam, et al., Multiple polynomial regression method for determination of biomedical optical properties from integrating sphere measurements, Appl. Opt. 39 (7) (2000) 1202–1209.
[4] J. Aspuru, A. Ochoa-Brust, R.A. Félix, W. Mata-López, L.J. Mena, R. Ostos, R. Martínez-Peláez, Segmentation of the ECG signal by means of a linear regression algorithm, Sensors 19 (2019) 775.
[5] R.O. Duda, P.E. Hart, Pattern Classification, John Wiley & Sons, 2006.
[6] K. Dramé, F. Mougin, G. Diallo, Large scale biomedical texts classification: a kNN and an ESA-based approaches, J. Biomed. Semant. 7 (1) (2016) 1–12.
[7] E. Frank, L. Trigg, G. Holmes, I.H. Witten, Naive Bayes for regression, Mach. Learn. 41 (1) (2000) 5–25.
[8] S.R. Safavian, D. Landgrebe, A survey of decision tree classifier methodology, IEEE Trans. Syst. Man Cybern. 21 (3) (1991) 660–674.

[9] O.I. Abiodun, A. Jantan, A.E. Omolara, K.V. Dada, N.A. Mohamed, H. Arshad, State-of-the-art in artificial neural network applications: a survey, Heliyon 4 (11) (2018), e00938.

[10] D.T. Pham, S.S. Dimov, C.D. Nguyen, Selection of K in K-means clustering, Proc. Inst. Mech. Eng. C J. Mech. Eng. Sci. 219 (1) (2005) 103–119.

[11] F. Murtagh, P. Contreras, Algorithms for hierarchical clustering: an overview, Wiley Interdiscip. Rev. Data Min. Knowl. Discov. 2 (1) (2012) 86–97.

[12] T. Ali, S. Asghar, N.A. Sajid, Critical analysis of DBSCAN variations, in: 2010 International Conference on Information and Emerging Technologies, IEEE, 2010, June, pp. 1–6.

[13] G. Chandrashekar, F. Sahin, A survey on feature selection methods, Comput. Electr. Eng. 40 (1) (2014) 16–28.

[14] S. Wold, K. Esbensen, P. Geladi, Principal component analysis, Chemom. Intell. Lab. Syst. 2 (1–3) (1987) 37–52.

[15] A. Giuliani, The application of principal component analysis to drug discovery and biomedical data, Drug Discov. Today 22 (7) (2017) 1069–1076.

[16] M.H. Hassoun, Fundamentals of Artificial Neural Networks, MIT Press, 1995.

[17] I. Stephen, Perceptron-based learning algorithms, IEEE Trans. Neural Netw. 50 (2) (1990) 179.

[18] S.H. Wang, P. Phillips, Y. Sui, B. Liu, M. Yang, H. Cheng, Classification of Alzheimer's disease based on eight-layer convolutional neural network with leaky rectified linear unit and max pooling, J. Med. Syst. 42 (5) (2018) 1–11.

[19] J.C. Bansal, Particle swarm optimization, in: Evolutionary and Swarm Intelligence Algorithms, Springer, Cham, 2019, pp. 11–23.

[20] J. Nobre, R.F. Neves, Combining principal component analysis, discrete wavelet transform and XGBoost to trade in the financial markets, Expert Syst. Appl. 125 (2019) 181–194.

[21] A.B. Stepanov, Wavelet analysis of compressed biomedical signals, in: 2017 20th Conference of Open Innovations Association (FRUCT), IEEE, 2017, April, pp. 434–440.

[22] A. Humeau-Heurtier, A.C.M. Omoto, L.E. Silva, Bi-dimensional multiscale entropy: relation with discrete Fourier transform and biomedical application, Comput. Biol. Med. 100 (2018) 36–40.

[23] X. Li, D.W. Tripe, C.B. Malone, Measuring Bank Risk: An Exploration of z-Score, 2017. Available at SSRN 2823946.

[24] U. Rajendra Acharya, P. Subbanna Bhat, S.S. Iyengar, A. Rao, S. Dua, Classification of heart rate data using artificial neural network and fuzzy equivalence relation, Pattern Recognit. 0031-3203, 36 (1) (2003) 61–68.

[25] T.J. Jun, H.M. Nguyen, D. Kang, D. Kim, D. Kim, Y.H. Kim, ECG arrhythmia classification using a 2-D convolutional neural network, arXiv preprint arXiv: 1804.06812 (2018).

[26] C.N. Gupta, R. Palaniappan, S. Swaminathan, S.M. Krishnan, Neural network classification of homomorphic segmented heart sounds, Applied Soft Comput. 7 (1) (2007) 286–297.

[27] N. Dey, A.S. Ashour, Ambient Intelligence in healthcare: a state-of-the-art, Glob. J. Comput. Sci. Technol. 17 (3–H) (2017).

[28] M. Elgendi, R. Fletcher, Y. Liang, N. Howard, N.H. Lovell, D. Abbott, K. Lim, R. Ward, The use of photoplethysmography for assessing hypertension, NPJ Digit. Med. 2 (1) (2019) 1–11.

[29] P.K.D. Pramanik, A. Nayyar, G. Pareek, WBAN: driving e-healthcare beyond telemedicine to remote health monitoring: architecture and protocols, in: Telemedicine Technologies, Academic Press, 2019, pp. 89–119.

CHAPTER 5

IoT and medical cyberphysical systems' road map

Contents

5.1 Introduction

The cyberphysical system is a paramount research domain in today's world. It helps us to design the ecosystem of the physical device that is controlled, monitored, and coordinated by an Internet-based communication infrastructure. There is wide use of the concept of CPS present in the modern world that includes healthcare, agriculture, defense, the aerospace industry as well as Industry and Society 4.0. To make a smooth road map in the modern healthcare system where patients and doctors can be connected within vast geographical distance, such a system helps to make a universal solution. Moreover, the recent pandemic situation also gives us a lesson that we should modernize the healthcare infrastructure so that the situation does not get worsen. This chapter emphasizes the recent issues and proposes some potential solutions in the field of medical cyberphysical systems and applications for smart healthcare applications.

5.2 Ubiquitous sensing paradigm

The ubiquitous sensing paradigm is an emerging sensing technology for sensor networks, body area networks, and many more. The majority of the industry and healthcare applications nowadays acquire the ubiquitous sensing paradigm for the development of smart sensing applications. Sensor networks are perhaps the powerful system to establish a distributed sensing infrastructure. This involves numerous computing and communication infrastructures. The advanced communication features like LTE, 5G, and 6G as well as the high-level cloud-based computing infrastructure like platform as a service (PaaS) and infrastructure as a service (IaaS) give a step forward to establish a highly dynamic and robust ubiquitous sensing infrastructure.

5.2.1 Internet of medical things

The demand for the wireless network is increasing day by day. Under this situation, the medical network is also expected to connect with the Internet. The 5G Internet technology evolved and is said to be a revolution in the field of low-latency networks. This guarantees an ultrahigh data rate with almost 5 Gbps. In recent years, KT Corporation also announced GiGA LTE that is supposed to be 10 times faster than the conventional LTE networks. Internet of Medical Things (IoMT) is, however, a most challenging technology that allows medical devices to be connected within a single network backhaul and provides an enormous and rapid data transfer between medical services and devices. IoMT provides services that majorly required a short amount of data with minimum latency, and the device also ensures a minimum amount of battery usage. The major data elements that the IoMT system supports are sensor data that are majorly in a bulk amount; image data that are either grayscale or RGB images with a high resolution; and video data for MRI, gait, and posture as well as animation. Also, some mission-critical applications must be considered which involves real-time vital sign information of the emergency medical observation of the critical patient. Fig. 5.1 shows an abstract implementation of the IoMT services that are connected with medical devices with the cloud- and fog-enabled system.

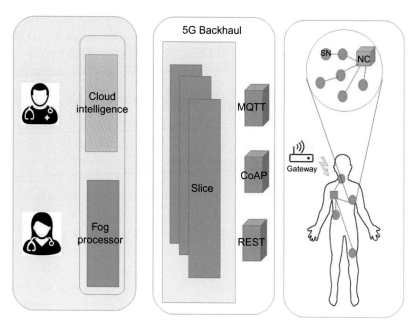

Fig. 5.1 Conceptual architecture of internet of medical things scenario.

The IoM devices [1] are also must be capable of the rapid amount of data and user information so that they are capable to perform the identification of the patients, doctors, and healthcare personnel. It also requires fast connectivity with the devices with a 5G network so that imaging and real-time sensor data gathering from the wearable and implantable devices are possible. In the next couple of subsections, we are discussing some futuristic methodology of the IoMT application that must be available within this decade.

(A) RFID-based patient management

This technology already came in several hospitals where patients are connected with their nearby hospital from home, and while in the starting of the diagnosis, the RFID-based authentication mechanism has to be considered in order to identify the patient identity and the status of the patient. The hospital doctors also have the same type of RFID mechanism to access the patient data and the pathological reports.

(B) Robotic system

Robots are nowadays taking leadership in various industrial works where there is a need for high-precision products. An activity like industrial welding and fault inspection is gaining popularity. In the medical domain also due to the advancement of 5G technology, there is a wide chance of introduction of the robotic systems. One of the best examples in this context is medical drones. These drones are capable of the delivery of the emergency medicine and medical equipment in remote and disaster-hit regions. Another big utility is remote robotic surgery. This concept is perhaps as game-changing as the introduction of the Tactile Internet. The research has been carried out in the field and IEEE 1918.1 standard has been come out to perform the remote Tactile activity with hepatic feedback through the low-latency Internet ecosystem.

(C) Wearable body sensor networks

Wearable and implantable body sensor networks are gaining popularity in today's world [2]. In medical systems, patient monitoring can be easily performed using the sensor that can be deployed on the body. There are two types of body sensors that are popular: (i) in-body sensors that are majorly deployed underneath the body itself and (ii) on-body sensors that are mainly deployed over the body for sensing the data like temperature, heart rate, body pH level from the sweat, and body glucose level. There is n number of biosensors that can be placed. Those biosensors generate data and transmit them toward the sinks for storage and processing at the edge level. The designing of WBAN involves a proper WSN topology formation and structure. The system should also ensure how the data are routed to the sink and further forwarded toward the cloud. In the case of data forwarding, typically wired medium is used. In another approach, the body itself is considered as a conductor and the message forwarding has been performed through a low-impedance body conductor. As such, in-body propagation involves high power consumption and message loss as well. While designing such a network, the system should ensure that no human tissue gets damaged due to overheating and electrical charge preservation which may result in uncomforting and health hazards to the patient.

Another major aspect of the BAN design ensures the routing methodologies that mainly ensure multipath routing. A mixed-integer programming design approach is highly preferable to generate a routing path in this case. Along with this, the energy-efficient protocol design is also considered as a crucial point in which multilayered topology design is highly preferred. In this multilayer approach, each layer deployed is considered as a set of clusters. A cluster head has to be selected based on the energy consumption by the node among all clusters. Finally, the cluster head collects the data from all clusters and forwards it toward the controller. All the clusters can communicate with the cluster of the above layer as well as the layer below it. There may be some link failure or anomaly that happens between the connectivity or due to the network disruption. To avoid such a scenario, the amplify-forward or decode-forward technique is highly suitable. Ample research has been carried out on sensor network design for body area networks for healthcare systems. The research is still carried out to achieve optimality in the context of energy routing interoperability.

5.2.2 Smart pill technology

Smart pills are mainly used for diagnosis. In recent days, however, they are widely used in various therapies and pathological treatments as well [3,4]. In the drug delivery system, smart pills are also having a significant role to play. The robotic capsules are such a class of smart pills that ensure a fully autonomous activity inside the body of a human being. There is a large class of applications of such robotic pills is present. One of them is the diagnosis and drug delivery in the intestine and large intestine. Personalized drug delivery is a crucial factor that ensures a proper dose suited for a specific patient according to the targeted location. The autonomous robotic systems in this case equipped with pH, temperature, and impedance sensors. The main job of these sensors is to manage the internal temperature, pH level, state of the internal organ while delivering the drugs. The main application of the drug delivery system (DDS) is to treat gastrointestinal disease, chronic inflammation of the large intestine, peptic ulcer, and gastrofungal diseases.

An ideal swallowable robotic pill must be fully autonomous and able to position and perform diagnosis. Such robots are connected with a receiver through a wireless connection. The typical dimension of such a capsule is 10 mm × 30 mm. The localization and power management are the key issues that should be incorporated into those sensor pills. The robot often comprises several passive sensors and the actuation devices like the liquid pump and small arm. To drive those devices, a minimum power requirement is necessary which is higher than the driving voltage of passive sensors.

Fig. 5.2A shows a component overview and Fig. 5.2B shows the actuation mechanism of a smart pill, respectively. The navigation of such capsules involves two different types of locomotion, namely active and passive. In the case of passive locomotion, the natural peristaltic contraction is the main locomotion factor of the capsules. This technique requires a minimum amount of energy consumption as the locomotion is caused by an external force. The disadvantage of this locomotion is the unreliable motion of the device causes an imprecise movement of the robotic capsule. The objective of the active locomotion is to perform a precise movement of the device to monitor and diagnose the internal anomaly. The active locomotion system must ensure to adhere to the tissue so that the locomotion force gets transmitted. Displacement of the contact point to generate the locomotion is necessary. In the case of liquid media, the robot must swim. In some cases, the internal locomotion of the robot has to be designed. In this case, the robot uses the earthworm or tapeworm principle. In such a case, the locomotion of the robot may be done with some hook or sucker-like mechanism through which the robot gets suspended inside the wall of the stomach and intestine without making harm to them. The communication mechanism has to be designed most efficiently to communicate with the ground station. In the majority of the robots, the HF and VHF frequency bands have been used because of their penetration capability. Some more sophisticated capsules also use RF telemetry. Nowadays, the wireless-, Zigbee-, and Bluetooth-enabled smart pills

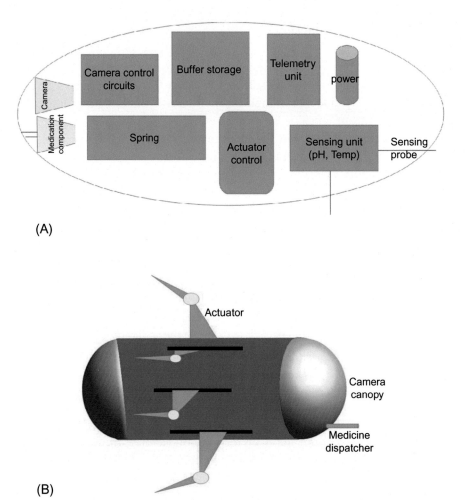

Fig. 5.2 (A) Basic structure of smart pill with the camera and medication component (B) smart pill with active locomotion actuator, camera, and medicine dispatcher module.

are also gaining popularity. The main bottleneck of such a communication mechanism is the frequency band used. As all of such communication is done at gigahertz-order frequency level, the penetration power of such signal is low; therefore, a significant amount of message loss has been observed in such cases.

The safety concern of the smart pill is also a crucial issue. While designing the smart pills, the safety must be ensured so that the contact of the capsule with the wall does not cause any damage to cells or tissues. In the case of earthworm locomotion, there is a high chance that the capsule was unable to avoid the damaged tissue area. On the contrary, the pills with legs and actuators allow better control in comparison with the normal earthworm that mainly offers passive control mechanisms. Another major advantage of legged motion is the precise maneuverability and safety. The legged pills involve the additional hardware involvement that has to be ensured a precise operation in terms of locomotion. Also, anchoring is a critical aspect of such legged pills.

5.3 IEEE 1918.1 tactile IoT

The low-latency paradigm of the 5G network system ensures a near real-time operation of the devices. This concept ultimately introduces a well-known Internet framework that is often known as Tactile Internet (TI). This concept has been standardized by the IEEE Tactile Internet standard working group which is popularly known as IEEE 1918.1 [5–7]. This concept of wireless communication is pushing the boundaries of the Internet application to the next level from home to remote physical interaction. It is primarily used to transfer audiovisual, hepatic, and kinesthetic (muscular movement). There are enough strict communication requirements that have to be followed during Tactile Internet communication. Along with this, ultralow-latency network connectivity has to be guaranteed. One of the fundamental objectives of such an ultralow-latency network is to control the sensorimotor. One of the popular examples of this technology is remote ball balancing and juggling. The balancing and the juggling of the ball involve the transfer of the data generated by the sensors deployed in the body of the human. The communication channel for TI must ensure minimum message loss to achieve the precise balancing and control with a mission-critical timeline. The situation is even more critical as the machine is a client instead of a human is taking part in hepatic interaction. This happens due to the extreme increment of the reactivity, impulsive force, and the enhanced capability of the mechanical devices. The latency of the TI system is also a crucial factor that is considered as a maximum benchmark of 5 ms of the round trip. The redesign of the communication network is a major goal in the development of the TI infrastructure so that the TI precise signal and the hepatic signal can be passed in a sophisticated way. The third-generation partnership project (3GPP) is working on that crucially to incorporate the latency reduction and reliability enhancement features. The fundamental goals of the Tactile Internet system are as follows:

(a) It provides a harness for remote physical interaction that is majorly required for exchanging action and hepatic reaction information at low latency.

(b) The information exchange is of different types but mainly they are either human to human, human to machine, machine to human, or a machine to machine as well.

(c) In the case of Tactile Internet methodology, the term object has various meanings. In some cases, the object can be termed as human, or any machine-like robots, network devices, routers or switches, protocols, and many more. The object must be a connected entity and they must do constant interaction with each other.

(d) Unlike software in the loop (STIL) or hardware in the loop, here the major concern is the human in the loop that performs physical interaction with hepatic feedback that is often known as bilateral hepatic teleoperation. In this mechanism, the machine and human both should not distinguish between local executable manipulation and remote exccutable manipulation.

(e) Machine-in-the-loop operation also must ensure that the operation and the interaction are in such a way that the system does not understand whether it is a local or remote operation.

(f) The hepatic information can be categorized into several different categories. The two major among them is tactile-kinesthetic or hybrid that comprises of both properties. The tactile information generally refers to perception information from various mechanical receptors such as texture, friction, skin temperature, and pH value. The kinetic information is received from human joints, bones, muscles, tendons, and many more.

5.4 Functional architecture

While designing the architecture of TI system, it is always considered to be a generic system. The architecture should consist of several modules. Each module must have interoperability capabilities so that the heterogeneous devices and network protocols. The network architecture also must ensure the proper QoS and QoE [8,9] of the network by managing the advanced functionality like lightweight signaling, caching, distributed computing framework, intelligent data analysis. IEEE P1918.1 supports the various modes of connectivity between one or more than one tactile device. The devices are often considered edge nodes because of the edge computing capabilities. In the majority of the cases, the tactile IoT nodes perform local-level processing and decision-making and buffering as well as catching. Each tactile edge consists of multiple TD or tactile devices. TD can communicate with various other devices and tactile edges. The communication networks in this case are majorly either wireless or wired networks. Sometimes they may be consisted of a dedicated wired or wireless network that forwards the message in low latency. Fundamentally, the architecture of the TI system comprises various components such as sensor node (SN), actuator node (AN), hepatic node (HN), and controller node (CN). The network gateway has to be controlled by a gateway network controller (GNC) under which there is a number of gateway networks (GNs) are available. The job of GN is to send the message and the control packet into the low-latency network backhaul. Under the network generally, "n" numbers of low-latency slices are working which are controlled through some dedicated access and core networks. Under network backhaul, the hard or soft slicing may apply to communicate to the remote object or hepatic nodes. The other tactile edges are also connected through the slice and received and responded accordingly. The architecture diagram of the TI systems is depicted in Fig. 5.3.

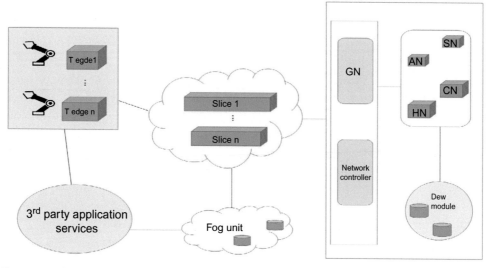

Fig. 5.3 Implementation of the tactile IoT ecosystem.

5.5 Applications and services

Tactile Internet surely performs a realistic social interaction with different devices. Since current WLAN infrastructure is not sufficient to achieve the end-to-end delay, this is a crucial factor for Tactile Internet. Some of the cutting-edge application of Tactile Internet related to 5G network infrastructure is discussed later.

5.5.1 Industrial automation

The machine-to-machine communication and machine-type communication are the key features of industrial automation. The main motive of such communication is to ensure a high data rate with low delivery latency and high response time. The Ethernet standard of today's world requires more low-latency operation and enhanced topology and routing standards to ensure ultralow-latency communication. Industrial IoT and industrial Ethernet can perform utmost responsibility to devices such as an ecosystem.

5.5.2 Robotics and motion planning

The diversified application of robotics has been implemented, which was designed in this era. The concept of aerial robotics, cloud robots, and network robots is the major game-changer. The autonomous and remotely controlled robots with visual hepatic feedback and synchronous control are taking a key part in numerous applications. The robotic systems which are dedicated to design for assembly lines for welding and assemble cars and automobiles must be run in a mission-critical scenario, and it is expected that the latency between the robotic system and the operator and its feedback must not exceed certain order of milliseconds.

5.5.3 Healthcare applications

The Tactile Internet is taking challenging steps in the healthcare and health informatics industry. Like telesurgery, telerehabilitation, and telediagnostic applications, this technology takes a step forward toward the next level of smart and critical healthcare tools. By using these tools, the medical consultancy can be available anywhere anytime mode. The telerobots in this case also controlled by the physicians. In this case, the information is not only audiovisual information but also hepatic information as well. In telerehabilitation, the motion and the control of the patient also can be steered remotely.

5.5.4 Augmented, virtual, and mixed reality applications

The advancement of the augmented, virtual, and mixed reality (AR, VR, and MR) is quite predominant in recent days. The existing AR, VR, and MR techniques are quite sufficient in a Tactile Internet application. The AR, VR, or MR application in this case has to share the hepatic virtual environment where the number of devices and users are physically connected via simulation and emulation tools. In contrast to static information augmented reality, the TI system must also be considered dynamic information AR for low latency and updated information. Hepatic feedback in Tactile Internet involves high interaction. Augmentation of the system also enables various assistance systems like driving assistance and navigation assistance; in the education sector also, such assistance systems are pretty widely used. Not only that in manufacturing, but entrainment sectors here are also a huge use of such AR, VR, and MR tools which are massively used. All these futuristic technologies are the forerunner of the Tactile Internet of things for sure.

5.6 5G and healthcare

The fifth-generation mobile communication (5G) platform is a game-changing technology that evolves to provide the utmost critical solution in the field of communication, Internet of Things (IoT), healthcare industry, and Business 4.0 [10,11]. Intelligent resource utilization is perhaps the most crucial factor in IoT-enabled smart healthcare systems. As the population of the world increasing rapidly, in recent times the infected diseases are also rapidly spreading throughout the globe. Therefore, the sophisticated management of the healthcare infrastructure with an effective communication infrastructure is highly required. Further, closely monitoring the infected diseases and the health conditions of elderly people in a frequent manner with a low-cost solution draws the attention. Not only that the employees in science technology, but administrative employees, teaching professionals, and health workers are prone to heart diseases due to work stress and the financial crunch, which leads to a serious physical disorder that must be monitored and diagnosed properly. Also in an emergency, immediate medication and monitoring of the vital signs are highly required. The multiprotocol platform is one of the tools that can be used as a specific IoT healthcare platform. Multiprotocol is a combination of various protocols accumulated together such as Zigbee, Bluetooth, LTE, MIMO as well as 802.11 b/g/n. This enables a heterogeneous infrastructure that ensures a wide and scalable end-to-end connectivity between the doctor, medical staff, and patients in an uninterrupted manner. This healthcare infrastructure has some fundamental components such as wireless body area networks, communication infrastructure, cloud storage, local-level processing and storage, and analytics tools.

The main job of the body sensor network, in this case, is to monitor blood pressure, oxygen levels, pulse rate, and respiration rate. The proposed ecosystem is a layered architecture that comprises a base layer that is the combination of the sensor network, smartphone, and lightweight computation device like Raspberry Pi. The sensor senses the physical quantity and sends it to the application of the smartphone or Raspberry Pi. The smartphone or the Raspberry Pi device in this case can act as an edge device. Some basic level of processing and the computation have to be done by such devices in this case. The data are then further redirected to the local computer. The physiological signal recognition has to be done in this through some optimized algorithms. In the second layer, the wireless communication infrastructure is present. This layer is comprised of the most sophisticated 5G network system that integrates the low-range wireless device with the fifth-generation infrastructure.

5.7 Wi-fi and the femtocell

The concept of femtocell is perhaps the extended concept of Wi-Fi technology that can be highly useful for communication and certainly replaces the wired networking concepts [12]. These networks are pretty much smart in the context of data offloading, connectivity, and latency of the network. As the 5G technology, we are dealing with millimeter waves; therefore, it is quite evident that the transmission range is also restricted to a very small distance as well as the coverage too. To develop a less expensive design, the femtocell concept is most useful; this model of cell design is also known as the home base station. The user can deploy the indoor base station for better coverage of voice as well as data. The major issues of such femtocell is that it affect the access point performance when it is deployed within an ultradense network. This leads to a saturated number of stations per access point. The cochannel interference limits the channel capacity and the allocation too. The dense access point deployment also restricts the channel's admission. The higher number of stations also results in compromising the throughput of the network due to the limitation of collision avoidance during the carrier sense multiple access.

A large amount of latency causes a link maintenance problem. Even in a high signal-to-noise ratio, the nonline-of-sight station works in a very low power which is at about 70–75 dBm. Frame management issues might be experienced due to CSMA/CA-style protocol. This leads to a high collision rate in a dense environment. To overcome such issues, client power control is highly recommended. While there is no client in the transmission range if we reduce the scanning interval, then the channel allocation, as well as the power consumption, can be reduced significantly. Another improvement feature is scalability where the network gets scattered through a large set of access nodes with proper interference management mechanisms. Secondly, mobility and roaming support are highly required which ensures uninterrupted connectivity of the mobile device and the AP with carrier-grade deployment. Further, the integration with radio access network (RAN) is also an utmost necessity. Here the backhaul Wi-Fi traffic mobile network systems must carry out billing, addressing, authentication, content filtering, roaming, and mobility management. Also, carrier security is another issue that the systems should be dealt with.

5.8 Software-defined networks

IoT-enabled infrastructure in healthcare and medical sensor networks is growing in leaps and bounds. The modern Internet of everything concept adds more devices to the network and it is expected to connect every hospital and the medical infrastructure to be connected in the network by 2030. As the number of devices increases in the IoT infrastructure, more dedicated bandwidths and resources are required for the service of such a huge number of devices. The 5G networking concept is developed by keeping that in mind. However, from the network point of view, the 5G backhaul infrastructure must ensure a high degree of Quality of Service (QoS). To achieve such QoS, the software-defined network (SDN) can be developed to ensure more devices can be served by using existing resources and the virtual infrastructure. The concept of SDN is not new, but it is the extension of the existing concept for the newly created low and ultralow-latency networks [13,14]. The SDN in this case is mainly created by applying network function virtualization (NFV) that ensures the functionalities, protocols, and resources to be virtualized in such a way that the physical resources. The NFV mechanism is a technique that intelligently performs the resource utilization and the load balancing of the network itself. The software that is dedicated to the NVF is generally run into the physical network devices like a switch, high-performance router, edge, or fog node in a network. In recent days, the NFV includes the network applications and the services as well. However, the major components of NFV are (1) the physical hardware elements like network gateway, NAT subsystem, and remote access servers for the broadband; (2) the mobile network devices like basic service set (BSS), home location register (HLR), and home subscriber server (HSS) and also the node mobility management system; (3) the virtualized home environment is another important component here; (4) tunneling, SSL, and VPN gateways; (5) traffic analysis unit and Quality of Experience (QoE) management frameworks; (6) and also, the majority of the part comprises traffic inspection elements like session controllers and multimedia subsystem controllers. We can consider that the fundamental advantage of the NFV is to virtualize the network environment that ensures the cost-effectiveness, reusability, and interoperability between different heterogeneous software and hardware platforms. In an NFV environment, the coexistence of different types of software and services is present which can perform the collaborative tasks. Fig. 5.4 illustrates the NFV mechanism.

The NFV mainly exploits the virtualized network functionalities (VNFs) which comprise commodity servers that are wrapped up in a layered architecture. This architecture supports logical partition between layers and services. VNF is mainly deployed in a virtual machine (VM) and managed by a management and orchestration system. In this case, the behavior of the system is driven mainly through metadata. There is a strong relation between SDN and the NFV. The concept of SDN and NVF is closely related and also they complement each other. NVF can be conceptualized as a key component of SDN that virtualizes the SDN controller as well. However, from an architecture and functionality point of view, the SDN and NVF are completely different. NFV's main goal is to softwarize the functionality of the network functions, whereas the main goal of SDN is to perform a centralized and programmable network control and architecture management. The fundamental objective of the NFV is to control the functionality of the network by tuning the space and the power consumption of the network, whereas the SDN provides

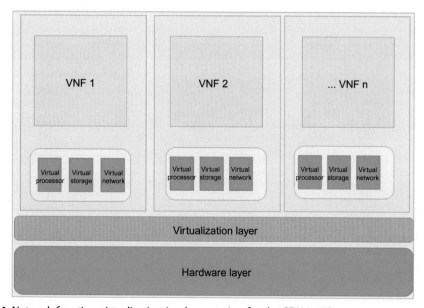

Fig. 5.4 Network function virtualization implementation for the SDN in 5G system.

flexible network control and abstraction. The NFV mainly emphasizes the software and hardware that are involved with the network itself. SDN on the contrary emphasizes the functionality of the network plane and the data plane and performs the coordination of the same. The amalgamation of such technologies results in the software-defined network function virtualization SD-NFV. It mainly comprises the control device and the NFV platform that executes at the edge of the network. Packet forwarding in this case is controlled by the SDN controller. They use the forwarding table to forward the packets to different destinations.

5.9 Slicing under SDN

Slicing is another key concept in a software-defined network. It implements network softwarization that ensures an end-to-end network running underneath a physical network. Each of the network slices uses physically common or separate hardware and software modules and services [15,16]. The slice instances in such cases are virtually isolated from each other. The slices are here independently controlled by the SDN controller. Fig. 5.5 shows the conceptual view of the network slices.

The network slices mainly comprise a set of resources that are precisely combined and met the service requirements. In general, three types of network resources are present in a network slice.

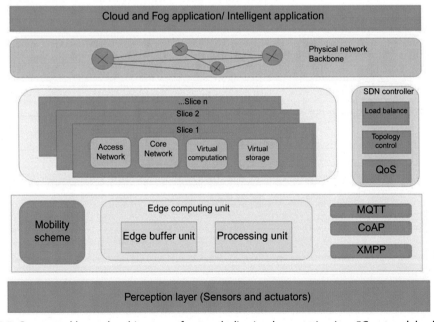

Fig. 5.5 Conceptual layered architecture of network slice implementation in a 5G network backhaul.

(i) Network function (NF): It consists of the functional blocks that are having dedicated networking capability. This can be deployed as a software instance that is running on infrastructure resources. The NF may be vendor-specific or sometimes open-source elements can also be considered as NF.

(ii) Resource for infrastructure (RI): These are commonly built by considering heterogeneous hardware infrastructure and modules. This hardware mainly consists of the network switch, router, and radio access network elements. To optimally use those resources, the attribute of the resources must be abstracted to leverage the virtualization mechanism.

(iii) Orchestration: This is a significant part of the network slicing. It signifies the correlation between the different software and hardware components of the slices. It also signifies the optimal utilization of the resources to fulfill the user requirements. It ensures client-specific service validation and event notification. The orchestration within network slices also ensures strong isolation. This is because, in general, the slices are dedicated to the specific service and each service set. The isolation must be understood based on certain criteria such as performance of the slice, data security in the slice resources, and management capability. The slice, in this case, is majorly defined for a particular service requirement known as a slice performance indicator. Moreover, the attacks and the threats are also pretty vital and important factors that should be addressed by the security module of the slice. Often when we deploy a network slice, there is a chance of accommodation of a huge amount of nodes, and as the number of node increase, the chances of security vulnerability are high. To achieve the proper isolation, the set of management policies has to be decided. The policies must be implemented in software, hardware as well as virtualization levels. To fully realize the proper isolation and virtualization, proper orchestration is recommended in the utmost way.

5.10 Drone as a component of healthcare and MCPS

UAV-based smart healthcare system is a challenging field of research in recent era. In several types of research, the researchers build several methodologies and algorithms to develop and deploy the drone service to manage emergency health services. Drone-based emergency heart patient assistance is one of the major application domains [17]. In this case, if the patients have a sudden heart attack, their device sends an automated message to the nearby health center as an SOS. The health center will immediately send the drone with minimum health support. The small UAV has been used with the inbuilt cardiopulmonary resuscitation (CPR) or chest compression system that can be used for emergency medical purposes.

Another major service that can be done using the drone is emergency medical supply in disaster-affected regions where it is difficult to supply food and medical aids. In most cases, those regions are majorly unreachable through normal paths due to the unavailability of the route. Although certain rules have been mentioned by the Federal Aviation Administration (FAA), Washington, nevertheless, the designer and the developers have to develop the UAV and the drone infrastructure by maintaining the following rules and the regulations for civilian drones as shown in Table 5.1.

5.11 UAV-assisted COVID-19 monitoring

The coronavirus disease 2019 (COVID-19) is one of the biggest threats in recent era throughout the globe. At least, 1.5 crores of new cases appear by today with 6.4 lakhs of death throughout the world. The severe acute respiratory syndrome coronavirus 2 (SARS-CoV-2) causes a severe respiratory syndrome and attacks the lungs. Fever, cough, and throat pain are the main symptoms in this case. Often, the body oxygen level gets drastically decreases in this situation. Although the sources are unknown, scientists have had discovered the genome sequence and found the similarity with the beta-CoV family. At the end of 2019, the Wuhan City in Hubei Province was first affected by the coronaviruses and at the beginning of 2020 started spreading out the whole globe. According to the World Health Organization (WHO), it causes a serious health problem. It is a spherical and mainly RNA-based virus ranging from 600 to 1400 Armstrong in diameter [18]. The clinical feature of the COVID-19 reported asymptomatic and acute respiratory distress syndromes. Also, multifunction dysfunction syndrome is reported in severe conditions.

Different stages of the corona outbreak have been observed in this scenario. In stage 1, the virus mainly spreads from the people who have visited or returning from those countries which are already affected. In such a situation, the virus outbreak mainly happens locally within a single cluster. In stage 2, the local transmission can happen when an affected person spared the virus to nearby people like family members, friends, and colleagues. In stage 3, the spreading is reported within a community. The cases in such a scenario are majorly the cluster of cases. The final stage is the transmission within a vast community and the disease spread out within a town or city massively.

Table 5.1 Standardized protocol for civil drone aviation mentioned by the FAA.

Parameters	Value
Weight of UAV	<25 kg (55 lbs)
Maximum speed	87 knots (100 kmph)
Maximum altitude	400 ft (above ground level)
Operational mode	Daytime only (with restricted night mode)
Working range	Preferably visible line of sight

5.12 The use of sensors and IoT infrastructure to fight against COVID-19

Digital emerging technologies play a pivotal role in to fight against the COVID-19. The major technologies like IoT, machine learning, and 5G communication networks are some major building blocks that are highly relevant in improving public health response. One of the biggest technologies that are to be considered in this scenario is the Internet of Medical Things which is the amalgamation of the medical network and the IoT infrastructure. Some of the IoT-based devices we consider in this scenario are as follows.

5.12.1 Smart thermometer

Digital thermometers are gaining popularity hugely due to the low-cost and high-precision reading. Some of the consumer electronics manufacturers are also developing thermometers that are connected to the Internet. Although these classes of thermometers are mainly developed to monitor flu patients in the United States, they are proved to be effective in fighting against corona. These thermometers are generally connected to the Internet either directly as they have a Wi-Fi interface in it or through the mobile app. Most of the thermometers in this case have low-cost Bluetooth or NFC modules so that they can be able to connect with the mobile device. The software framework of the application then processes the temperature data and generates statistical trends of the fever within a selected province. This is perhaps extremely helpful to monitor and track the increment of the COVID-19 cases within the hot spots and containment zones.

5.12.2 Networks and cloud robots

There is a huge demand for robotic devices in the pandemic situation. As the COVID-19 is a highly infectious disease, some of the major uses of robots are in the COVID ward of the hospital. We can use wheeled robots to carry the food and necessary medicines to the COVID patients. As well as the robots can be used as an assistant of the doctor during a check-up of the COVID patient that can carry necessary types of medical equipment. These robots can also be treated as part of the Internet so that the real-time location can be tracked. Such robots are also sometimes operated remotely as well. The robots can also be used to carry the seriously ill patients from the entrance of the hospital to the wards. Due to the advancement of IoT and cloud technology, the robots can also connect and share information with the cloud platform. In such a case, the robot motion control and motion path planning data streamed into the cloud. The cloud platform further processes the data received from the robot unit itself, and finally, the decision has been made and the response sent back to the robot to instruct the robot to perform a certain task.

5.12.3 Autonomous vehicles

Under the pandemic situation, the monitoring of the containment zone and the detection of the COVID-positive patient within the congested area are pretty evident. The drone and the ground vehicle-based containment zone monitoring are pretty common in this scenario. In such a case, the movement of the people under the containment zone gets strictly monitored. Even in some cases, the drones might consist of thermal and infrared cameras. By using those camera units, it is possible to get the body temperature information directly. Due to the virtue of edge computing, even we can analyze such images with onboard lightweight edge computation units and decide the body temperature of the people in a mass. Not only that by analyzing the body temperature data obtained from the thermal scan but even the prediction and forecasting of the potential chance of the spreading of COVID-19 can be possible too.

By using the recent advanced Internet of Things and Internet of Everything infrastructure, it is quite possible to perform the analysis of the data in the remote server, and based on the result generated, the decision can be made.

5.13 Drone delivery in COVID-19 situation

In a pandemic situation like COVID-19, the major concern is the social distancing. Drones in such cases can be considered as a primary requirement for major and critical transportations. The people are constantly advised to avoid congested and highly trafficked areas. Drones are therefore can be utilized as grocery carriers, health equipment carriers, and emergency medicine carries as well. To achieve these goals, proper path planning and mobility modeling are needed. During this situation, the various initiatives on drone-based mobility and communication by various public and private-sector agencies have been reported [19]. In the United States, the FAA tries to make a separate drone highway so that the transportation of critical goods becomes simpler. This will facilitate the rural area and food desert a key service of the delivery of the groceries. In this way, the continuous supply chain can be made from the groceries and pharmacy warehouses which are mainly situated in metropolitan areas to the remote rural area.

Fig. 5.6 shows the implementation of the drone network in a lockdown scenario. Three stages have been categorized. In stage 1, the COVID-19 fighter and doctor are doing constant monitoring and treatment on the patient or potential patient. The devices like the digital thermometer and testing kits are considered as sensor nodes that are gathered the data. These data are then forwarded to the base station by considering the drone as a data forwarder node which lay in stage 2. These drones are mainly used to grab the sensor data through wireless, LTE, or 5G channel. The drones are also capable to perform edge-level processing of the data gathered by them. Such a drone ecosystem has to be deployed in various containment zones and hospital localities to monitor and gather

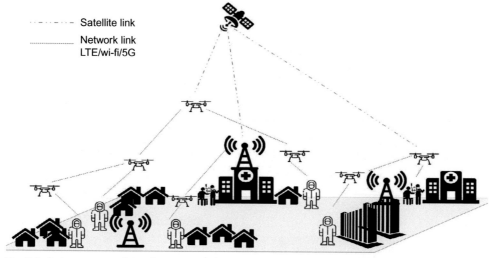

Satellite link
Network link
LTE/wi-fi/5G

Fig. 5.6 A drone network implementation for COVID-19 monitoring and information gathering.

the data. Therefore, there must be a link between different zone clusters so that they can interact and share data. To perform data transfer within such a cluster, either long-range LTE or satellite communication is preferred.

5.14 COVID-19 prediction modeling

The COVID-19 is a highly infectious disease. This often causes massive illness, organ failure, and death. The prediction of the infection within a certain community is a major challenge to avoid the community spread. There are several mathematical models through which we can predict the possibility of the potential spread of the COVID-19 infection. One of the simplest and most popular models among them is the susceptibility, infected, recover/remove model which is often called as S-I-R model [16,20]. The model has been established with a basis of the total population of a geographical region as T_p which is a constant value and we can express it by $T_p = S + I + R$. In this context, we should consider two major factors: (i) the total population of the area is constant for the ideal case and (ii) the rate of infective population is proportionally constant with the susceptible population. Also, another assumption we have to make is that the recovery/death rate is also constant concerning the infected rate. Therefore, the model can be realized by governing a differential equation model as depicted in Fig. 5.7.

We can consider the rate of change of susceptibility $s(t)$ as follows:

$$\frac{ds}{dt} = -r \cdot I \cdot S \qquad (5.1)$$

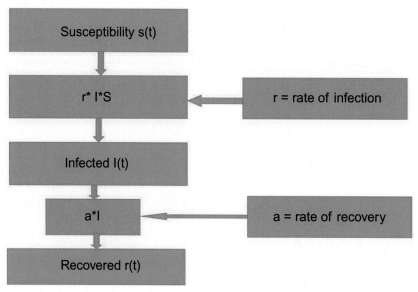

Fig. 5.7 Diagrammatical representation of the S-I-R model.

where $s(t)$ is computed concerning time, r is the constant rate of infection spread, I is the infected population value, and S is the susceptible population value. Here it is observable that the value of $\frac{ds}{dt}$ is majorly a negative that signifies that the susceptibility population is gradually decreasing. By applying the same rule, the rate of change of infected population can be computed as follows:

$$\frac{dI}{dt} = r \cdot I \cdot S - a \cdot I \qquad (5.2)$$

This signifies the gain in the infected category as the value of $\frac{ds}{dt}$ is negative so it is quite obvious that the $\frac{dI}{dt}$ value is positive that signifies a gain in the infected category. Even though there is a gain, nevertheless, a loss in the infected category has been added with a factor of $-a.I$. This signifies the loss in infected people which gives the positive weight to the recovered or removed category. Thus, the rate of change of the recovery can be computed as follows:

$$\frac{dR}{dt} = a \cdot I \qquad (5.3)$$

where "a" is the recovery rate constant. Therefore, the aforementioned linear differential equations are the illustration of the susceptibility, infected, and recovery rate of the population. The susceptibility rate is a sharply decreased rate that signifies that the majority of the people have a chance to infect as the number of infected people increases. The graph also shows as time increases, the rate of recovery has an exponential growth that signifies a positive result in terms of recoverability. The infected rate shows a spike nature whose first phase is an exponential increment of the infected population, and after a certain time, it reaches global maxima. After that, the infected population decreases rapidly due to increasing recovered or diseased people or due to hard immunity. To realize the contact ratio and the basic reproductive number, some major assumption has to be made as follows:

The total initial susceptible population $S = S_0$; an initial number of the infected population $I = I_0$; and the initial number of recovered population $R = 0$. Here each value is constant. If we assume total population $N = S + I + R$, then the initial population $N_0 = S_0 + I_0$ (as $R = 0$). Now, if we perform differentiation concerning t, then $\frac{dN_0}{dt} = 0$ which signifies the total population. Now from the relation, it is obvious that if I_0 increases, then the infected population increases. And, the value of S_0 is very large which naturally signifies the value of $S_0 \geq S$ always. So if we put this relation in Eq. (5.2) then.

$$\frac{dI}{dt} < r \cdot I \cdot S_0 - a \cdot I \quad \text{so,} \quad \frac{dI}{dt} < I(r \cdot S_0 - a) \tag{5.4}$$

Now from the relation mentioned previously, it is clear that if $(r \cdot S_0 - a) > 0$, then the disease will spread rapidly. This relation also signifies $S_0 = \frac{a}{r}$ which can be represented as a new term known as $\frac{1}{q}$ which is also known as the contact ratio. This signifies a function that represents the number of people who come into contact with infected people. Now from here, the basic reproduction number R_0 is the number of secondary cases in which one case produces a completely susceptible population. It majorly depends upon the period of the infection. A higher value of R_0 signifies a greater chance of pandemic and epidemic causes due to infection. The value of R_0, in this case, can be considered as $R_0 = \frac{rS_0}{a} > 1$. The condition $R_0 > 1$ is crucial here which is signifies the chance of massive spread of the infection.

5.15 Conclusion

The application of IoT in healthcare has been discussed in this chapter. The chapter mainly focused on the utilization of various networking technologies like wireless body area, in-body, and on-body sensor networks. Also, based on the recent technology like 5G and the conceptual representation of tactile IoT, network slices and software-defined networks along with network function virtualization have also been addressed. Finally, the role of IoT and IoMT in COVID-19 monitoring and detection along with the latest advancement of COVID-19 drone technology has been addressed.

References

[1] S. Rani, S.H. Ahmed, R. Talwar, J. Malhotra, H. Song, IoMT: a reliable cross layer protocol for internet of multimedia things, IEEE Internet Things J. 4 (3) (2017) 832–839.

[2] S. Goswami, P. Roy, N. Dey, S. Chakraborty, Wireless body area networks combined with mobile cloud computing in healthcare: a survey, in: Classification and Clustering in Biomedical Signal Processing, IGI Global, 2016, pp. 388–402.

[3] A. Koulaouzidis, D.K. Iakovidis, A. Karargyris, E. Rondonotti, Wireless endoscopy in 2020: will it still be a capsule? World J Gastroenterol: WJG 21 (17) (2015) 5119.

[4] C.M. Klugman, L.B. Dunn, J. Schwartz, I. Glenn Cohen, The ethics of smart pills and self-acting devices: autonomy, truth-telling, and trust at the dawn of digital medicine, Am. J. Bioeth. 18 (9) (2018) 38–47.

[5] O. Holland, R. Eckehard Steinbach, V. Prasad, Q. Liu, Z. Dawy, A. Aijaz, N. Pappas, et al. (Eds.), The ieee 1918.1 "tactile internet" standards working group and its standards, Proc. IEEE 107 (2) (2019) 256–279.

[6] M. Dohler, 5G networks, haptic codecs, and the operating theatre, in: Digital Surgery, Springer, Cham, 2020, pp. 71–86.

[7] X. Wei, Q. Duan, L. Zhou, A QoE-driven tactile internet architecture for smart city, IEEE Netw. 34 (1) (2019) 130–136.

[8] N. Gholipoor, H. Saeedi, N. Mokari, E.A. Jorswieck, E2E QoS guarantee for the tactile internet via joint NFV and radio resource allocation, IEEE Trans. Netw. Serv. Manag. 17 (3) (2020) 1788–1804.

[9] M. Aazam, K.A. Harras, S. Zeadally, Fog computing for 5G tactile industrial internet of things: QoE-aware resource allocation model, IEEE Trans. Ind. Inf. 15 (5) (2019) 3085–3092.

[10] A.A. Barakabitze, A. Ahmad, R. Mijumbi, A. Hines, 5G network slicing using SDN and NFV: a survey of taxonomy, architectures and future challenges, Comput. Netw. 167 (2020) 106984.

[11] Q.-V. Pham, F. Fang, V.N. Ha, M.J. Piran, M. Le, L.B. Le, W.-J. Hwang, Z. Ding, A survey of multi-access edge computing in 5G and beyond: fundamentals, technology integration, and state-of-the-art, IEEE Access 8 (2020) 116974–117017.

[12] I. Budhiraja, S. Tyagi, S. Tanwar, N. Kumar, M. Guizani, Cross layer NOMA interference mitigation for femtocell users in 5G environment, IEEE Trans. Veh. Technol. 68 (5) (2019) 4721–4733.

[13] D. Camps-Mur, J. Gutierrez, E. Grass, A. Tzanakaki, P. Flegkas, K. Choumas, D. Giatsios, et al., 5G-XHaul: a novel wireless-optical SDN transport network to support joint 5G backhaul and fronthaul services, IEEE Commun. Mag. 57 (7) (2019) 99–105.

[14] W. Zhuang, Q. Ye, F. Lyu, N. Cheng, J. Ren, SDN/NFV-empowered future IoV with enhanced communication, computing, and caching, Proc. IEEE 108 (2) (2019) 274–291.

[15] D.A. Chekired, M.A. Togou, L. Khoukhi, A. Ksentini, 5G-slicing-enabled scalable SDN core network: toward an ultra-low latency of autonomous driving service, IEEE J. Sel. Areas Commun. 37 (8) (2019) 1769–1782.

[16] L.B. Azzouz, I. Jamai, SDN, slicing, and NFV paradigms for a smart home: a comprehensive survey, Trans. Emerg. Telecommun. Technol. 30 (10) (2019), e3744.

[17] I. Martinez, A.S. Hafid, A. Jarray, Design, resource management and evaluation of fog computing systems: a survey, IEEE Internet Things J. 8 (4) (2021) 2494–2516.

[18] R. Sabino-Silva, A.C.G. Jardim, W.L. Siqueira, Coronavirus COVID-19 impacts to dentistry and potential salivary diagnosis, Clin. Oral Investig. 24 (4) (2020) 1619–1621.

[19] R. Estrada, M. Arturo, The Uses of Drones in Case of Massive Epidemics Contagious Diseases Relief Humanitarian Aid: Wuhan-COVID-19 Crisis (Available at SSRN 3546547), 2020.

[20] I. Cooper, A. Mondal, C.G. Antonopoulos, A SIR model assumption for the spread of COVID-19 in different communities, Chaos, Solitons Fractals 139 (2020), 110057.

CHAPTER 6

Smart perishable food and medicine management overview

Contents

6.1. Introduction

In due course of time, it can be stated that the temperature, moisture, and some other different parameters of the food and medicine is one of the most important parameters which is required to be well maintained and also taken good care of, especially in the field of food supply chain management. Some of the foods and medicines are extremely sensitive to those parameters. Thus during the making of these foods as well as medicines and also during packaging, the temperature needs to be very carefully monitored. Refrigeration is one of the most effectively used methods for the maintenance of proper temperature. Proper control and maintenance of the medicine and food temperature are very important while delivering perishable foods and lifesaving medicine to the consumer. The implementation of IoT can be of good help in such a process. This chapter mainly emphasizes the review of the application of IoT in the control and maintenance of the quality of critical commodities in the field of food and medical supply chain management. Further, a hybrid vehicular delay-tolerant network-based IoT approach is addressed and

Biomedical Sensors and Smart Sensing
https://doi.org/10.1016/B978-0-12-822856-2.00001-0

analyzed so that effective quality management can be done during the transportation of food and medicine.

At first, when the Internet of Things was thought of getting implemented in the field of the medicine supply chain, the focus was that the concept of IoT will be used to integrate every aspect of the food supply chain starting from farm to fork and medicine from factory to pharmacy. To avoid the damage of crops, farmers are now implementing the concept of precision agriculture. Precision agriculture is a farming method that mainly considers the variability of soils, pests, and crop yields depending on which portion of a field is getting worked upon. Fields vary in terms of soil types, moisture content, contour, and crop yield, how we plant corn over an entire area will also have to vary depending on our location in the field. After the farming state, the food and medicine reach the logistic stage which can include warehousing, distribution, and delivery to retailers. At this stage, the focus is on what the temperature and humidity conditions are concerning the products. At the distribution level, the foods and medicines are transported via trucks, rail, or plane. Here IoT can play a major role. It tracks the geographical progress of the food to determine the fastest and safest route to the market. It continuously monitors the environment of the interiors of refrigerated trailers and the interiors of packages and containers that meats and produce are stored. If the seal of a container is broken, or temperature and humidity controls within the container fail, the sensors issue immediate alerts to supply chain managers so the situation can be mitigated. Collectively, these food tracks, trace, and control mechanisms reduce spoilage and maintain the track and trace of foods from farm to table.

In addition to that, the food and medicine supply chain tangibly depends upon the food and medicine transportation from sources like cold storage of farmhouse to the distributor and the retail shop. To manage the quality of food and medicine chain management, the IoT plays a vital role. In such a scenario, the medicine and foods may be distributed within a vast geographical location from centralized storage. This involves good transport network support. Nowadays airfreights are the most convenient way for the transportation of food and healthcare equipment. Along with that, a specialized network structure can also be thought of. The introduction of IoT in such a scenario is crucial. With IoT, intervehicular communication is the primary necessity to monitor and track the status of medicine and foods. To make it possible a vehicular ad hoc network with a delay-tolerant network can be designed which collectively gathers the information of the medicine and food quality within the network itself. This work primarily emphasizes a review of the current trends of the application of the Internet of Things in medicine and food hybrid supply chain management. Some state-of-the-art extension of the existing IoT-based food and medicine management has been addressed. The extension has been thought of as an ecosystem that comprises a hybrid architecture that consists of a vehicular DTN-based IoT system. In this case, each network node is acted as a sensor node to track the status. The node further applies flooding or forwarding methods to update each other's status. Some of the stationary nodes among them act as IoT gateway from where the food and medicine quality status has been broadcast to the cloud server. The motivation of the work, therefore, emphasized a hybrid model of vehicular DTN-based IoT architecture.

Also in earlier days, medicine management can be performed using simple electrical circuitry. In some cases, lifesaving drugs and vaccines need absolute precise temperature management. The classical LED-based system has been incorporated in medicine management in various rural and urban healthcare centers and hospitals. Nowadays the introduction of smart sensors ensures more precise management. We can consider some IoT-based ecosystems for medicine management. The system must comprise multiple sensors like ambient temperature, ambient light intensity, humidity, and pH level. In the majority of cases, humidity plays a vital role in medicine management. The calcium and zinc-based drugs often reduced their activities in an extremely humid situation. Special care must be taken to those drugs which are highly prone to damage.

6.2. Food and medical supply chain perspective

6.2.1 Food supply chain

A food supply chain is a process that determines how the foods reach our houses—the process which the food undergoes from the stage of its farming to our tables. The processes include production, processing, distribution, consumption, and disposal. The food that we eat reaches us after undergoing a food supply chain process. Every step of the supply chain requires human and/or natural resources. In the food supply chain, food moves from producer to consumer via the processes of production, processing, distribution, retailing, and consumption. At the same time, the money that consumers pay for food moves from consumers to producers in the reverse process, again from consumer to retailer to distributor to processor to farmer.

A typical supply chain begins with the ecological, biological, and political regulation of natural resources, which are then followed by the human extraction of raw material. Maintenance of food quality is of the utmost importance in this management system. Out of many parameters, the maintenance of the proper temperature and humidity plays a major role. The time gap or the time elapsed during the intermediate processes of the food supply chain management system can affect the initial condition of the food. Thus the condition of the food can deteriorate. Thus to maintain such a condition before it reaches the consumer end, the application of the concept of the Internet of Things can be of great use.

6.2.2 Medicine supply chain management

In an integrated healthcare system, the medical supply chain plays a crucial role. The manufacturer of the medicine has two very specific pathways to supply medicine to the patient.

In the first case, the medicine can be supplied through the pharmacy and from it delivers to the clinic. Later the patients who are visiting that clinic got the medicine. In the second case, the medicine gets routed to the hospital and then it gets routed to the hospital pharmacy. Along with the medicine, the medicine information flow is also a crucial thing that should be addressed. The medicine supply chain can be modeled in three steps. The multi-objective and multiperiod supplier selection methodology has been considered a crucial aspect. The integer programming methodology can be considered an important phenomenon. Survey reveals that supply chain cost is a crucial factor that needs to be addressed. The advancement of the computational system and advanced algorithm surely ensures multi-objective cost optimization. The fuzzy model is more suitable in this case. To realize multi-objective and multiperiod supply chain problems, interactive fuzzy programming has been introduced that specifies the satisfaction of the level of each objective. This optimization methodology ensures the optimal value for the decision problem. The accuracy of the method lies in constrained satisfaction value while achieving the desired objective.

6.3. Internet of things concepts

Internet of things can be imagined as a "global neural network." IoT includes some smart devices which interact with each other. The systems with which they are interacting are like machines, environments, objects, and infrastructures and the Radio Frequency Identification (RFID). Communication capability and remote manual control lead to another step that emphasizes how can things be automated and, based on some particular settings and with sophisticated cloud-based processing, how things can happen without interventions. That is the ultimate goal of some IoT applications. But to successfully incorporate the Io concept the devices have to have an embedded processor-based controlling system. There are few layers in the architecture of the Internet of Things. The first layer is the smart devices like the sensors and actuators, the next layer is the sensor data acquisition system called the edge gateway and the sensor data preprocessing part, and the last layer includes the cloud application built for IoT using the microservices architecture, which is usually polyglot and inherently secure using HTTPS/OAuth. The application areas of the IoT are smart homes, elder care, medical and healthcare, transportation, building and home automation, manufacturing, agriculture, etc.

MQTT, MQTT-SN, RTMP, and RTSP protocols can be used highly to transfer the messages from supply vehicles or specific centers like pharmacies, food storage, etc. MQTT in this case runs in three different Quality of Service (QoS) values, namely, QoS0, QoS1, and QoS2. In QoS1, the protocol must ensure at least one message delivery wherein QoS2 ensures exactly 1 message delivery. QoS0, on the other hand, doesn't guarantee message delivery. In the case of MQTT-SN, only we can use QoS0 and QoS1. Other Quality of Services are not supported by MQTT-SN due to their lightweight nature. RTMP protocol, on the other hand, can be used for real-time audio and video streaming. This protocol was developed by Macromedia. One of the major advantages of this protocol is the low latency. This protocol is also adaptable so users can rewind the feed as per requirement. The RTMP is also a flexible protocol that can support numerous formats like m3a, Flv, aac, mp4, etc. and there are some disadvantages. RTMP does not support HTML5 format. It is also vulnerable to low bandwidth which can seriously affect the performance of the streaming. Also, HTTP and HTTPS support is not present. RTSP, on the other hand, is a less popular protocol and it is based on the streaming server. The live stream in this case can be controlled by using play, pause, and record commands. The big advantage of the same is the segmented streaming and customized protocol support like TCP and UDP protocols. The bandwidth performance of RTSP is better than RTMP and the latency is comparatively low in this case. HTTP and HTTPS are also not supportable by this protocol.

6.4. Literature survey

It is observed that compared with the other product supply chains the most difficult one is the food and medicine supply chain. The reason is that food is perishable and every food has its shelf life. In the case of medicine, high temperature may cause medicine damaged. The cold chain method is the most used. From the sensitivity of the food point of view, the temperature is the most prominent parameter and foods become too much time and temperature sensitive at times. Thus each step, namely harvesting, preparation, packaging, transportation, and handling in the entire chain, has to be taken care of. The paper [1] Temperature management for the quality assurance of a perishable food supply chain, Myo Min Aung, Yoon Seok Chang, Food Control 40 (2014), focuses on the methods of determining the optimum temperature in case of multicommodity refrigerated storage. It also experiments with Wireless Sensor Network. This work concludes that this method of quality monitoring and assessment in a real-time platform is much superior compared to the conventional visual assessment method. Also, it shows the advantages of Euclidean distance cost depending on temperature changes.

The paper [2] Food Traceability: New Trends and Recent Advances. A Review, Badia-Melis R., Mishra P., and Ruiz-García L., Food Control (2015), emphasizes the record maintenance of the food. Here it is shown that the RFID method can be used to increase the wheat flour sale. The smartphone can also be used to know the total IV range of a food product. Analysis of the DNA sequence or knowing the food authenticity based on some isotopic analysis can also be used for proper authentication of the food supply chain. Food traceability can also be determined by using the Internet of Things like the development of a common framework toward unifying the present technical regulations, the interconnectivity between agents, environment loggers, and products. Internet of Things can be used as intelligent traceability of the temperature of the food product and its remaining shelf life.

The paper [3] Internet of things and supply chain management: a literature review, Ben-Daya Mohamed, Hassini Elkafi, and Bahroun Zied (2017), International Journal of

Production Research, is a literature review on the effectiveness of the Internet of Things on the food supply chain management. It reviews many aspects of the Internet of Things like methodology, industry sector, and classification. A bibliometric analysis of the literature is also presented. Areas of future SCM research where IoT implementation can be used are also identified.

The paper [4] A study on the decision-making of food supply chain based on big data, Ji Guojun, Hu Limei, and Tan Kim Hua (2016), shows how the concept of big data harvesting can be utilized to maintain the food based on market feedback in the food supply chain. This can be utilized to make the supply chain system more and more informative. Here the Bayesian network is applied to predict the demand of the market based on the integration of sampled data and the cause–effect relationship. A deduction graph model is used here. At first, the demand is translated into the process and further the process is divided into tasks and assets. It was concluded that this analytical model can be of great use in this food supply chain.

In the work [5], The "Intelligent Container"—A Cognitive Sensor Network for Transport Management, Lang Walter, Jedermann Reiner, Mrugala Damian, Jabbari Amir, Krieg-Brückner Bernd, and Schill Kerstin, (2011), suggests a sensor network called "Intelligent container." This sensor network manages the logistic process which specially deals with perishable foods like fruits and vegetables. The temperature and humidity are the two main parameters that are mainly monitored. To estimate temperature-related quality losses, detect malfunctioning sensors, and control the sensor density and measurement intervals, the cognitive system can make use of several algorithms. Decision support tools help in decentralized decision making which is based on the knowledge base of the goods, their history, and context. Thus the intervention of the headquarters of the logistics company and the container is reduced. The quality of process control is enhanced. Also, self-evaluation is possible depending on the sensor data.

The paper [6] The internet of things: a survey, Li Shancang, Da Xu Li, and Zhao Shanshan (2014), is a survey paper that shows the interventions of IoT in almost all aspects of today's world, especially when we have entered the era of smart technology in our daily life needs too. IoT encircles smart communicating things. This concept leads to a time where the world will be filled with physical entities and virtual components. In this survey, the definitions, architecture, fundamental technologies, and applications of IoT are sequentially reviewed. The work in this paper has four major parts. The first part introduces various definitions of IoT, then secondly it elaborates the different applications of IoT, thirdly it focuses on the issues related to the implementation of IoT, and finally the major challenges which are needed to be addressed by the research community and corresponding potential solutions.

In the paper [7] Social media data analytics to improve supply chain management in food industries, Singh Akshit, Shukla Nagesh, and Mishra Nishikant, (2016), a big data analytics approach is proposed considering the data acquired from social media like Twitter, to identify the supply chain management issues in food industries. Particularly a

Support vector machine and hierarchical clustering with multiscale bootstrap resampling algorithm is used for the text analysis. The cluster of words obtained as the result of this approach informs about customer feedback and issues in the flow/quality of food products to the supply chain (SC) decision-makers. A case study in the beef supply chain was analyzed using the proposed approach, where three weeks of data from Twitter were used.

The work in the paper [8] Wireless sensor network for real-time perishable food supply chain management, Wang Junyu, Wang He, He Jie, Li Lulu, Shen Meigen, Tan Xi, Min Hao, and Zheng Lirong (2014), shows the development of a real-time perishable food supply chain monitoring system based on ZigBee-standard wireless sensor network (WSN). To fulfill those requirements some important improvements like a configurable architecture for comprehensive sensors and an improved network switching scheme are designed. A tree-topology WSN system with 192 End Devices and a star-topology WSN system with 80 End Devices are implemented and evaluated in terms of both functions and performance. The success rate of this approach is about 99% according to the end results.

In the paper [9] Food safety prewarning system based on data mining for a sustainable food supply chain, Wang Jing and Yue Huili (2016), a food safety prewarning system is proposed based on association rule mining and Internet of Things. The main purpose is to timely monitor the data detected from the whole supply chain and automatically prewarn the food supply chain system. A case study on the dairy producer was conducted. From the case study, it was clearly shown that the proposed system can very well identify the risks and accordingly decide whether a warning should be issued.

The paper [10] IOT-enabled Quality Management Process Innovation and Analytics in China's Dairy Industry: A Data Flow Modeling Perspective, Han Yangyang, Feng Yuqiang, Liu Luning, J. Jingrui, and Wang Zhanfeng (2015), analyzes the then situation of the supply chain of the diary system of China. Then they proposed the IoT-based quality management approach. Then a business process modeling approach based on data flow perspective to describe this innovation management mode was employed to further emphasize the importance of the concept of IoT. The design approach was mainly based on the matrix of dairy products and process information.

The paper [11] Internet-of-Things paradigm in food supply chains control and Management, Riccardo Accorsi, Marco Bortolini, Giulia Baruffaldi, Francesco Pilati, and Emilio Ferrari (2017), encircles the motives and the path to follow to design and build IoT-based architecture for the planning and management of the food supply chain. It systematically gives the architecture of the entities; the physical objects; the physical and informative flows; the stages; and the processes to be sensed, tracked, controlled, and interconnected which elaborates the correlation between the observed supply chain and the exogenous environment.

The book [12] Food Supply Chain Management and Logistics, Dani Samir, elaborates on the different steps of the food supply management system comprising of the steps followed from the farming of the food to the foods delivered to the consumer end. From the farming, the steps followed are manufacturing of the food which faces some operational challenges and also has the logistic end to it. Also during the manufacturing, the points which are needed to be considered are future sustainability and challenges. Then it goes to the retailing end and then finally to the consumers.

The book [13] Fundamentals of supply chain management, Lu Dr. Dawei (2011), mainly emphasizes the different aspects of Supply Chain Management. The points covered are requirements of supply chain management and how this management works globally. Then it also focuses on the planning and designing of this chain system. Then it leads to the purchasing and supplier selection.

The paper [14] Application of IoT in Supply Chain Management of Agricultural products, Shambulingappa H.S. and Pavankumar D. (2017), states that Supply Chain Management (SCM) is the most predominant part of today's global industries. SCM is the sequence of the flow of goods and services. SCM comprises of movement and storage of raw materials and goods from one place to another place. It is composed of a design, planning, execution, control, and monitoring of goods. It controls the product flow, information flow, and financial flow. Agricultural supply chain traceability should be established to enhance the development of the internet of things. In this system, the farmer gets complete information on the whole life cycle of the product. Thus transparency is maintained by optimizing the supply chain. Connecting them and their product directly in the SCM can be of great advantage for the farmer community.

The work [15] Application of Internet of Things in food packaging and transportation, Maksimović Mirjana, Vujović Vladimir, and Omanović-Mikličanin Enisa (2015), states that food safety is such a scientific field that includes many routine steps and inspection at each stage. These steps are mainly followed to avoid dangerous health risks. Food quality can be monitored at any point from farm to table, connecting at the same time food producers, transportation, and hospitality/retail companies with the help of internet of things (IoT) connected testing equipment. In this paper mainly IoT applications in food packaging and transportation are demonstrated. A low-cost solution based on IoT is also proposed.

In the survey journal [16] Food supply chain management: systems, implementations, and future research, Zhong Ray, Xu Xun, and Wang Lihui (2016), there is a review on systems and implementations of food supply chain management system. A systematic and hierarchical framework is proposed in this paper to review the literature. Almost 192 papers are reviewed in this survey. The papers are related to data-driven systems for FSCM. Due to the large demand for implementations of automation in supply chain management systems, IoT-based supply chain management monitoring is in high demand.

The journal [17] Implementation of Food Quality Tracking System Based on Internet of Things, Feng Tianzhong (2016), states that the management of food quality which is based on the internet of things is an integrated monitoring and management information system. It consists of intelligent database technology, radio frequency identification technology, food safety technology, multimedia and network technology, wired and wireless network technology, and other practical high-tech techniques. According to this journal, the system is a combination of detection of food safety, tracking its quality, and then network-based integration. GSM /GPRS public wireless network for remote data transfer is used by the proposed system. It proposes that the combination of GSM/GPRS public wireless network and the internet of things can also be useful in cost reduction.

The paper [18] Internet of Things Applications on Supply Chain Management, B. Cortés, A. Boza, D. Pérez, and L. Cuenca (2015), proposes the use of IoT in the field of supply chain management. It has been seen that nowadays the concept of IoT has been used in many industrial applications. Sensing Enterprise (SE) is a network that allows it to react to business stimuli originating on the Internet. Some evidence has also been discovered on the use of IoT in supply chain management.

The research paper [19] Supply Chain Management and Sustainability in Agri-Food System: Italian Evidence, Zecca Francesco, and Rastorgueva Natalia (2014), shows that supply chain has become predominant in agriculture in recent times. A conceptual framework of the contemporary agricultural supply chain processes is proposed in this paper. Various influencing factors are examined and a few issues according to the Triple Bottom Line concept are also considered. The logistics part of the supply chain was paid the maximum attention to the work. It is stated that sustainable logistics has an economic, social, and environmental impact on rural areas. The agricultural value chain was described based on the analysis of Italian evidence based on different indicators.

The journal paper [20] Virtualization of food supply chains with the internet of things, C.N. Verdouw, J. Wolfert, A.J.M. Beulens, and A. Rialland (2015), shows that in the operational management system the virtualization of the supply chain management is a very important development. Virtualization enables supply chain actors to monitor, control and plan, and optimize business processes remotely and in real time through the Internet. The virtual food supply chain management system is analyzed from the IoT point of view.

The vehicular ad hoc network is another major concern in food supply chain management because several numbers of food supply trucks have been deployed to supply the food by food chains. To manage the synchronization of the status of the food within the supply change system, the formation and the message transfer between the vehicles are a major concern. To achieve such a solution some of the operations can be addressed in this context. An intelligent VANET transportation system has been addressed for the logistics has been discussed by Khaliq et al. [21]. This concept encompasses the car-to-car cooperation model that provides information about the customized goods as well as it is used for the entertainment purpose of the passenger as well. The book "E-Logistics: Managing Your Digital Supply Chains for Competitive Advantage" by Wang et al. [22] gives the direction toward the modern food supply chain. The main emphasis on the food supply chain management, in that case, is the implementation of the airfreights. Chen et al. [23]

propose a unique approach to adaptive food mileage service inference based on vehicular network communication. In this task, real-time traffic information collection and allocation and transportation for food mileage service on the information system have been proposed. Linear programming, *4D Routing Path Algorithm* (4DRPthe A), *Break the Law of Enclosing* (BLE), compute the average speed within each path. In this way the faster and shorter routing path to transport food. Thibaud et al. [24] in a work have addressed the applications of IoT in high-risk environments like intelligent transportation, food supply chain, and the healthcare industry. In this work, the various aspects of food supply chain management and its different bottleneck have been addressed. The issues like perishability and variability of the product, sustainable requirement, the need for the information system, and lack of feedback support system were the primary focus. The concept of connected vehicles has also been illustrated where the main point is the improved transportation ecosystem and the need for low latency and the robust real-time application has been addressed.

The management of the medical product is also utmost crucial. During the transport of medicine and other lifesaving drugs, often major issues and damages happen due to improper transport methodology, jerking, temperature, excessive humidity, and excessive light. The proper care of such medical equipment must be taken to avoid such things. Also, scheduling of the medicine and the preservation at home is a very crucial thing that needs to be addressed. Li et al. [25] describe the implementation of a medicine box that can serve as a medical care unit. This unit takes care of the temperature of the medicine. It preserves the optimal temperature so that the quality of the medicine can persist. Zhang et al. [26] propose a self-management tool for type-1 diabetes. Here an iterative user-centered design cycle has been addressed in this case. An app called MyDay has been deployed where the initial data entry has been performed. Also, the app provides detailed medical information about the patient. In another work da Silva et al. [27] propose a smart medicine box that supports scalability and support for medicine manipulation. The proposed system incorporates the edge computing concept into the medicine box. The entire status of the medicine in this case gets sensed and sent to the cloud server using MQTT publish/subscribe methodology. A medicine RFID management system has been proposed by Shieh et al. [28]. The proposed methodology ensures the work balance for those hospital members who are having a heavy workload. RFID is a crucial technology to develop ubiquitous systems. In this application, the author mainly uses a 13.56 MHz frequency for identification. In this methodology when the doctor prescribes some medicine to the patient they also assign a unique ID to the patient. When the nurse comes to the patient she checks the RFID of the patient and retrieves all medical information. The value is then sent to the medical store through the Internet (Table 6.1).

Table 6.1 Comparative study of the previous works related to IoT-based food quality management

Sl. no.	Title of the work	The focus point of the work
1	Temperature management for the quality assurance of a perishable food supply chain	The methods of determining the optimum temperature in the case of multicommodity refrigerated storage
2	Food traceability: New trends and recent advances. A review	RFID method for the record maintenance of the food
3	Internet of things and supply chain management: A literature review	A literature review on the effectiveness of the Internet of Things on the food supply chain management
4	A study on the decision-making of food supply chain based on big data	The concept of big data harvesting can be utilized to maintain the food based on market feedback in the food supply chain
5	The "Intelligent container"—A cognitive sensor network for transport management	A sensor network called "Intelligent container" manages the logistic process which specially deals with perishable foods like fruits and vegetables
6	The internet of things: A survey	The interventions of IoT in almost all aspects of today's world
7	Social media data analytics to improve supply chain management in food industries	A big data analytics approach considering the data acquired from social media like Twitter, to identify the supply chain management issues in food industries
8	Wireless sensor network for real-time perishable food supply chain management	The development of a real-time perishable food supply chain monitoring system based on ZigBee-standard wireless sensor network (WSN)
9	Food safety prewarning system based on data mining for a sustainable food supply chain	A food safety prewarning system based on association rule mining and internet of things
10	IoT-enabled quality management process innovation and analytics in china's dairy industry: A data flow modeling perspective	The situation of the supply chain of the diary system of China in 2015
11	Internet-of-things paradigm in food supply chains control and management	The motives and the path to follow to design and build IoT-based architecture for the planning and management of the food supply chain
12	Food supply chain management and logistics	The different steps of the food supply management system comprising of the steps followed from the farming of the food to the foods delivered to the consumer end
13	Fundamentals of supply chain management	The different aspects of supply chain management

Continued

Table 6.1 Comparative study of the previous works related to IoT-based food quality management—cont'd

Sl. no.	Title of the work	The focus point of the work
14	Application of IoT in supply chain management of agricultural products	Supply chain management (SCM) is the most predominant part of today's global industries
15	Application of internet of things in food packaging and transportation	Food safety is such a scientific field that includes many routine steps and inspections at each stage
16	Food supply chain management: systems, implementations, and future research	A review on systems and implementations of food supply chain management system
17	Implementation of food quality tracking system based on internet of things	The management of food quality which is based on the internet of things is an integrated monitoring and management information system
18	Internet of things applications on supply chain management	The use of IoT in the field of supply chain management
19	Supply chain management and sustainability in agri-food system: Italian evidence	A conceptual framework of the contemporary agricultural supply chain processes
20	Virtualization of food supply chains with the internet of things	In the operational management system, the virtualization of the supply chain management is a very important development
21	Synergies of advanced technologies and role of VANET in logistics and transportation	An intelligent VANET transportation system
22	Towards fog-driven IoT eHealth: promises and challenges of IoT in medicine and healthcare	RFID-based medicine management
23	A design of IoT-based medicine case for the multiuser medication management using a drone in an elderly center	Medicine reminder system using sensor and actuator
24	Medicine reminder and monitoring system for secure health using IoT	IoT- and RFID-based monitoring system implementation
25	Using IoT technologies to develop a low-cost smart medicine box	Implementation of Fog-based IoT architecture for medicine data acquisition and transfer
26	RFID medicine management system	Medicine information has been embedded in RFID Tag which can be acquired using a smartphone

6.5. The ecosystem of the IoT-based system

As we know in the case of IoT-based devices all the ecosystem (Fig. 6.1) is connected with a distributed environment. The fundamental parameter of the food quality which is addressed in this case has to be in the primary focus. As we know, the fundamental cause of food poisoning is rotten food. This happens due to bacteria and viruses. Depending on the bacteria and virus severity of the food will also increase. Protozoa roundworm is some of the parasites that are also the potential cause of food poisoning. While conceptualizing IoT-based food safety the sole target of the methodologies should be focused on identifying bacterial and fungus-based poisoning. The work has been done where the ambient conditions have been monitored at every stage of the supply chain. In such a case, the sensor will be deployed within the container itself. The specific sensor grid-like mechanism can be introduced which can be further implemented as a subsystem. The subsystem can be considered as a layered approach which is of having three layers. The base layers are the sensor comprised of the multisensor network. This layer has a series of temperature sensors like the LM35, humidity sensor, air quality sensor, and ambient light sensor. The 2nd layer is a communication layer that is having a communication protocol like IEEE 802.11 b/g/n or IEEE 802.15. This layer is also having some minimum processing power so that the sensor data can be arranged in a set of meaningful information. The communication channel performs the primary task in IoT because the node should broadcast the information from the source to the gateway. The gateway, in this case, is supposed to be the third layer. The IoT infrastructure [29] is the final layer that visualizes the data. Numerous processing algorithms have been incorporated into this

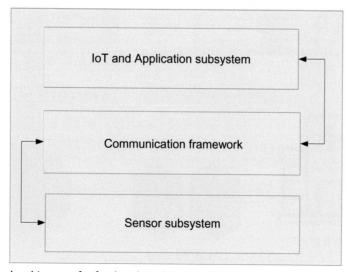

Fig. 6.1 Layered architecture for food and medicine quality monitoring of IoT system.

layer. The sole task of this layer is not only visualizing the data but also analyzing and performing decisions on the status of the food quality. Fig. 6.1 shows the layered architecture of the IoT ecosystem.

6.5.1 IoT solution for medicine

The concept of IoT nowadays gets applied in various sectors along with healthcare and supply chain management. Medicine management is also a crucial factor in this scenario. The people with regular work-life are considered being healthier. If the work cycle of the people hampers or is not in the proper form, then the wellness of the normal people is in an abnormal state. This tracking method is helpful for those people who are willing to live an independent lifestyle. But most older people are prone to an accident in such a situation. Various systems of monitoring elderly people have been developed. The main objective of such systems is to track the activity of the elderly people inside the smart home and respond accordingly. The response can be of various types. In type 1, the scheduled medicine alarm can ring if elderly people forgot to take the medicine in time. An intelligent medicine box is one of the potential solutions for this (Fig. 6.2).

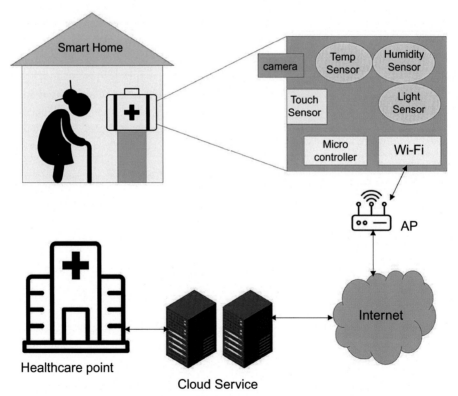

Fig. 6.2 Medicine box alert system for an elderly patient.

Such medicine box monitors medicine status, such as medicine quantity, moisture, and the temperature of medicine. Also excessive light can damage medicine sometimes. So the light sensor is also a crucial component in this case. The medicine box also tracks the touch so that it senses how many times it is utilized by elderly people. Also, two camera modules can be utilized as part of the medicine box to monitor the quantity of the medicine and also the activity of the people present in the home.

The sensors are then connected with a microcontroller where the data are gathered. The edge level computation has also been done in this layer. The fused data is then sent to the cloud server and analysis has been made. The medicine box in this case is connected to the Internet through a Wi-Fi access point. The video and the image information are generally sent using the RTSP protocol with a frame rate of 30 fps. The analytics engine in the cloud server consists of an object detection module which is mainly designed with TensorFlow or Pytorch technology.

6.5.2 Hybrid vehicular DTN-based IoT methodology

The complex supply chain network can also be designed in such a way that the components like vehicles could share real-time information about the quality of the food. In such a scenario, the IoT perspective can be thought of as a level that is perhaps merged with a vehicular ad hoc network-like scenario. In this case, each node can be considered as a VANET [30] node that is also treated as a fundamental component of the IoT base station. The nodes also have their routing protocol so that the information interchange has been performed by them. The abstract architecture of the same is depicted in Fig. 6.3.

Since the architecture has been a hybrid of vehicular ad hoc networks and the IoT, it exploits the properties of both the architecture. Not only that the system can be further

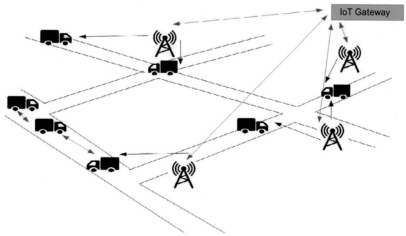

Fig. 6.3 Vehicular DTN-based IoT architecture for the food and medicine supply chain.

modified in the form of a Delay-Tolerant Network [31,32] in that case the system has to be considered as a set of Sparse nodes which perform an opportunistic routing among several nodes as they make a contact with each other. This will increases the chance of the updated message delivery. In such a case, some of the nodes can be the stationary node that hosts or broadcasts the crucial status information. It has been considered in this case the network topology in such a situation will always be dynamic. In such a case, opportunistic flooding and forwarding have to be performed. We can consider some of the routing methodologies for such sparse vehicular network scenarios.

6.5.3 DTN-based IoT approach

Since the DTN approach is an emerging technique for data forwarding among sparsely connected networks hence we can consider the DTN-oriented interconnection to update the status of each vehicle in a wide geographical region. We can consider following DTN protocols for the standard map-based movement model [33,34], in this case the vehicular nodes are supposed to choose the shortest path from the source to the destination. The protocols have been discussed as follows.

6.5.3.1 Direct contact method

In this case, the message exchange has been done as two vehicles come in direct contact with each other. It can be categories as flooding as well as a forwarding mechanism. In the case of flooding, the set of relays only contains destinations. In the case of forwarding, the relay will select the direct path between sources to destination [35]. This method is primarily applicable for VANET for the food supply truck that will make e contact rang of any base station or stationary nodes. This approach may increase throughput but waiting time may increase.

6.5.3.2 Epidemic routing methods

This mechanism ensures a sufficient amount of random exchange of data in a certain frame of time. The strategy is good for the highest delivery chance [36,37] of the data packet as several data exchange is high but the number of a message generated during the exchange is extremely huge. The message that has been sent from the base station must have a unique ID. When two nodes contact they exchange the messages and ensures that all nodes got the message of the same ID. This routing is not very complex because to execute it no knowledge of the entire network is required.

6.5.3.3 Location-based routing mechanism

This mechanism fundamentally depends upon geographical location information. The approach is promising and increases the efficiency of the data packets. Based upon the geo-location of the destination the source will send the data packet in this case. This approach avoids the forwarding of the packets using the network address. Like geographic routing [38,39] this method is also aware of the geographical location of the node itself as well as the other nodes. Since this method is a restricted forwarding technique so it is an efficient technique in terms of energy efficiency.

6.6. Analysis of different methodologies

To analyze the scenario of the hybrid vehicular DTN-based IoT ecosystem, we consider the efficiency of the above-mentioned routing in DTN which has been implemented to derive the state-of-the-art vehicular DTN-based IoT ecosystem. We have chosen those methodologies because of their performance, easy to deploy feature, and quality of service. To achieve the fundamental performance metric, we have chosen to evaluate delivery ratio, buffer usage, and quality of service as crucial parameters. The analysis has been done based on the transportation time of the food and medicine supply chain from source to destination. We have simulated the scenarios in an opportunistic network simulator to evaluate the hybrid vehicular DTN-based IoT model performance.

6.6.1 Message delivery ratio comparison

To compare the message delivery ratio between three different echo systems with three different routing methodologies, we have observed (Fig. 6.4) that the message delivery ratio for the epidemic is initially outperforming among all strategy but after the time elapse the delivery ratio slightly decreases due to the increment of the number of the buffer. Since this is a rapid flooding method so after a certain amount of time the buffer may get exhausted and the delivery of the message gets delayed. On the other hand, location-based routing starts with a slow nature of delivery but as time increases the delivery ratio of the network significantly increases. This is due to the nature of the geographical location base forwarding where the precise location of the destination node has been

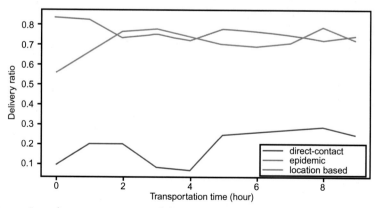

Fig. 6.4 Comparison between message delivery ratio within the specified time for a given methodology.

observed and the message forwarded more precisely to the destination node. The direct contact technique, on the other hand, performs not so well in comparison to location-based and epidemic routing. This is because in the case of direct contact the opportunity and chance to get the node in contact between two nodes is very less.

6.6.2 Buffer usage comparison

The buffer usage comparison between these three methods is depicted in Fig. 6.5.

In this case, the buffer used for the different routing strategies has been addressed. The epidemic routing, in this case, suffers from high buffer usage because of the rapid flooding nature. On the other hand, the location-based method has a constant decrement of the buffer usage up to the time of 8 hours. After that, a slight increment of the buffer usage has been observed. In the case of direct contact routing, on the other hand, buffer usage significantly achieves the lowest benchmark which is a minimum of 2 MB and a maximum of 3 MB. This is due to the nature of the strategy where a very small copy of the message has been generated during routing.

6.6.3 Comparison of quality of services

The comparison of the Quality of Service has been illustrated as follows (Fig. 6.6).

In this case, we can easily observe that the QoS is decreasing for all the strategies. Since the amount of time elapses the quality of the food and medicine will also get in a challenged situation. In the case of the direct contact model, this condition is most significant because of the unavailability of network information throughout the network. In such a case, the IoT station assumes the failure of the vehicular nodes which is perhaps the core reason for the decreasing QoS. On the other hand, location-based routing will ensure better services among all because the location of the vehicular node has been monitored at each point in time. In the case of epidemic strategy, the QoS falls drastically due to the

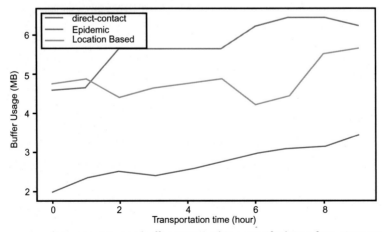

Fig. 6.5 Comparison between message buffer usage within a specified time for a given methodology.

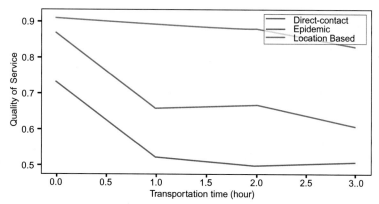

Fig. 6.6 Comparison between QoS within a specified time for a given methodology.

more amount of duplicated messages in the network which might create congestion in the network so the updated message may not be delivered within a certain time threshold to the IoT station.

6.7. Conclusion

The unique strategy of the vehicular DTN-based IoT framework for food and medicine management has been discussed in this chapter. We have found a location-based vehicular DTN approach for IoT-based food and medicine management which is significantly feasible and reveals a better solution for the food and medical supply chain. Food and medicine quality management, which is the most crucial approach, has been addressed and the different existing phenomena are illustrated in the literature survey. Further, the system can be implemented in a real-life scenario for real-time monitoring of the food and medicine quality of the entire supply chain. There is further research scope present for such a scenario where the energy management of the IoT-based vehicular DTN for food and medicine supply chains must be presented. Also, vehicular and flying networks such as airfreights can be integrated to improve the service of the food supply chain management and emergency medicine supply chain management.

References

[1] M.M. Aung, Y.S. Chang, Temperature management for the quality assurance of a perishable food sup-ply chain, Food Control 40 (2014) 198–207.
[2] R. Badia-Melis, P. Mishra, L. Ruiz-García, Food traceability: new trends and recent advances. A review, Food Control 57 (2015) 393–401.
[3] B.-D. Mohamed, H. Elkafi, B. Zied, Internet of things and supply chain management: a literature review, Int. J. Prod. Res. (2017), https://doi.org/10.1080/00207543.2017.1402140. ISSN: 0020-7543.
[4] G. Ji, H. Limei, K.H. Tan, A study on decision-making of food supply chain based on big data, J. Syst. Sci. Syst. Eng. 26 (2) (2017) 183–198.

[5] L. Walter, J. Reiner, M. Damian, J. Amir, K.-B. Bernd, S. Kerstin, The "intelligent container"—a cognitive sensor network for transport management, IEEE Sens. J. 11 (3) (2011).

[6] L. Shancang, D.X. Li, Z. Shanshan, The Internet of Things: A Survey, Springer Science + Business Media, New York, 2014, https://doi.org/10.1007/s10796-014-9492-7.

[7] S. Akshit, S. Nagesh, M. Nishikant, Social media data analytics to improve supply chain management in food industries, Transp. Res. E (2016), https://doi.org/10.1016/j.tre.2017.05.00. ISSN-1366-5545/_2017, Elsevier.

[8] W. Junyu, W. He, H. Jie, L. Lulu, S. Meigen, T. Xi, M. Hao, Z. Lirong, Wireless sensor network for real-time perishable food supply chain management, Comput. Electron. Agric. 110 (2014) 196–207. (2015) ISSN-0168-1699/_2014 Elsevier https://doi.org/10.1016/j.compag.2014.11.009.

[9] W. Jing, Y. Huili, Food safety pre-warning system based on data mining for a sustainable food supply chain, Food Control (2016), https://doi.org/10.1016/j.foodcont.2016.09.048. (2016), Reference: JFCO 5297.

[10] H. Yangyang, F. Yuqiang, L. Luning, J. Jingrui, W. Zhanfeng, IOT-enabled quality management process innovation and analytics in china's dairy industry: a data flow modeling perspective, in: Wuhan International Conference on e-Business 2015 Proceedings, aisel.aisnet.org/whiceb2015/23, 2015.

[11] R. Accorsi, M. Bortolini, G. Baruffaldi, F. Pilati, E. Ferrari, Internet-of-things paradigm in food supply chains control and management, Procedia Manuf. 11 (2017) 889–895.

[12] D. Samir, Food Supply Chain Management and Logistics, 2015, ISBN: 978-0-7494-7364-8.

[13] L.D. Dawei, Fundamentals of Supply Chain Management, 2011, ISBN: 978-0-7494-7364-8.

[14] H.S. Shambulingappa, D. Pavankumar, Application of IOT in supply chain management of agricultural products, Int. J. Eng. Dev. Res. 5 (3) (2017). ISSN: 2321-9939.

[15] M. Mirjana, V. Vladimir, O.-M. Enisa, Application of internet of things in food packaging and transportation, Int. J. Sustain. Agric. Manag. Inform. 1 (4) (2015), https://doi.org/10.1504/IJSAMI.2015.075053.

[16] Z. Ray, X. Xun, W. Lihui, Food supply chain management: systems, implementations, and future research, Ind. Manag. Data Syst. 117 (9) (2016) 2085–2114. 2017, Emerald Publishing Limited, ISSN-0263-5577 https://doi.org/10.1108/IMDS-09-2016-0391.

[17] F. Tianzhong, Implementation of food quality tracking system based on internet of things, J. Chem. Pharm. Res. 8 (4) (2016) 436–443. ISSN: 0975-7384.

[18] B. Cortés, A. Boza, D. Pérez, L. Cuenca, Internet of things applications on supply chain management, World Acad. Sci. Eng. Technol. Int. J. Comput. Inform. Eng. 9 (12) (2015). International Science Index waset.org/Publication/10003163.

[19] Z. Francesco, R. Natalia, Supply chain management and sustainability in agri-food system: Italian evidence, J. Nutr. Ecol. Food Res. 2 (2014) 20–28. ISSN-2326-4225, 2014 American Scientific Publishers https://doi.org/10.1166/jnef.2014.1057.

[20] C.N. Verdouw, J. Wolfert, A.J.M. Beulens, A. Rialland, Virtualization of food supply chains with the internet of things, J. Food Eng. (2015) 1–9. ISSN-0260-8774 https://doi.org/10.1016/j.jfoodeng.2015.11.009.

[21] K.A. Khaliq, A. Qayyum, J. Pannek, Synergies of advanced technologies and role of VANET in logistics and transportation, Int. J. Adv. Comput. Sci. Appl. (IJACSA) 7 (11) (2016) 359–369.

[22] Y. Wang, S. Pettit (Eds.), E-Logistics: Managing Your Digital Supply Chains for Competitive Advantage, Kogan Page Publishers, 2016.

[23] C.-H. Chen, H.-Y. Kung, C.I. Wu, D.-Y. Cheng, C.-C. Lo, The adaptable food mileage services inference based on vehicular communication, Adv. Sci. Lett. 5 (1) (2012) 101–108.

[24] M. Thibaud, H. Chi, W. Zhou, S. Piramuthu, Internet of things (IoT) in high-risk environment, health and safety (EHS) industries: a comprehensive review, Decis. Support. Syst. 108 (2018) 79–95.

[25] J. Li, W.W. Goh, N.Z. Jhanjhi, A design of IoT-based medicine case for the multi-user medication management using drone in elderly centre, J. Eng. Sci. Technol. 16 (2) (2021) 1145–1166.

[26] P. Zhang, D. Schmidt, J. White, S. Mulvaney, Towards precision behavioral medicine with IoT: iterative design and optimization of a self-management tool for type 1 diabetes, in: 2018 IEEE International Conference on Healthcare Informatics (ICHI), IEEE, 2018, pp. 64–74.

[27] D.V. da Silva, T.G. Gonçalves, P.F. Pires, Using IoT technologies to develop a low-cost smart medicine box, in: Anais Estendidos do XXV Simpósio Brasileiro de Sistemas Multimídia e Web. SBC, 2019, pp. 97–101.

[28] H.-L. Shieh, S.-F. Lin, W.-S. Chang, RFID medicine management system, in: 2012 International Conference on Machine Learning and Cybernetics, vol. 5, IEEE, 2012, pp. 1890–1894.

[29] E.N. Ganesh, Implementable approach of addressing the challenges of cloud based smart city using IoT, Rev. Energy Technol. Policy Res. 3 (2) (2017) 40–50.

[30] H. Hasrouny, A.E. Samhat, C. Bassil, A. Laouiti, VANet security challenges and solutions: a survey, Veh. Commun. 7 (2017) 7–20.

[31] S. Patra, A. Balaji, S. Saha, A. Mukherjee, S. Nandi, A qualitative survey on unicast routing algorithms in delay tolerant networks, in: Information Technology and Mobile Communication, Springer, Berlin, Heidelberg, 2011, pp. 291–296.

[32] S. Saha, A. Sheldekar, A. Mukherjee, S. Nandi, Post disaster management using delay tolerant network, in: Recent Trends in Wireless and Mobile Networks, Springer, Berlin, Heidelberg, 2011, pp. 170–184.

[33] D. Hrabcak, M. Matis, L. Dobos, J. Papaj, Social based mobility model with metrics for evaluation of social behaviour in mobility models for MANET-DTN networks, Adv. Electr. Electron. Eng. 15 (4) (2017) 606–612.

[34] A. Mukherjee, N. Dey, N. Kausar, A.S. Ashour, R. Taiar, A.E. Hassanien, A disaster management specific mobility model for flying ad-hoc network, Int. J. Rough Sets Data Anal. (IJRSDA) 3 (3) (2016) 72–103.

[35] M. Lamhamdi, Study of the energy performance of DTN protocols, in: 2017 Intelligent Systems and Computer Vision (ISCV), IEEE, 2017, pp. 1–7.

[36] Y. Sun, Y. Liao, K. Zhao, C. He, Performance evaluation of dtn routing protocols in vehicular network environment, in: International Conference on Machine Learning and Intelligent Communications, Springer, Cham, 2017, pp. 666–674.

[37] N. Kawabata, Y. Yamasaki, H. Ohsaki, On message delivery delay of epidemic DTN routing with broadcasting ACKs, in: 2017 IEEE 41st Annual Computer Software and Applications Conference (COMPSAC), IEEE, 2017, pp. 701–704.

[38] S.V. Alone, R.S. Mangrulkar, Implementation on geographical location based energy efficient direction restricted routing in delay tolerant network, in: Power, Automation and Communication (INPAC), 2014 International Conference on, IEEE, 2014, pp. 129–135.

[39] N. Benamar, K.D. Singh, M. Benamar, D. El Ouadghiri, J.-M. Bonnin, Routing protocols in vehicular delay tolerant networks: a comprehensive survey, Comput. Commun. 48 (2014) 141–158.

CHAPTER 7

Overview of data gathering and cloud computing in healthcare

Contents

7.1 Introduction

Wearable technologies [1] have changed the very way in which continuous monitoring can be carried out. The technologies developed have greatly impacted the well-being of individuals. The sensors are developed to gather various physiological changes about the wearer and analyze the information further. The basic physiological sensing is heart rate, temperature, etc. which our modern-day smartwatches can calculate. Researches on gait analysis for analyzing various disease conditions like Parkinson's, arthritis, etc. are also done with the help of inertial sensors like accelerometer and gyroscopes. In the present-day research domain, for the progression of the healthcare industry, data gathering is a very important aspect. The data gathered from the individuals need to be routed toward the sink node for further analysis. Cloud computing also plays a very important role in the present day, as advanced shared storage and computational analysis are required concerning the increasing dimension of data about the medical and clinical community.

Biomedical Sensors and Smart Sensing
https://doi.org/10.1016/B978-0-12-822856-2.00005-8
131

7.2 Wireless sensor network

Wireless sensor network (WSN) [2] forms the basis for supporting ubiquitous healthcare frameworks. With an increasing range of applications in the domain of medical application, the collection of data is done using various microelectronic sensors attached to one's body. The data from the wireless sensors are routed toward sink nodes or servers wirelessly in a hop by hop manner. Previously, the healthcare systems developed focused on hospitals and various clinics. But in the present era, the monitoring of health data has reached person's home and in our day-to-day life. Smart homes are built on the pretext of a wireless sensor network that employs intelligent heath monitoring of the residents. The embedded sensors sense the users or inhabitants and form an intelligent monitoring framework. Thus the collection of data is very important for both sharing and analysis. A wireless sensor network can be defined as the framework where hundreds and thousands of infrastructure-less sensors are deployed to collect and monitor various conditions, in this case, health conditions, and routes the data toward a sink node or to any other node in the network.

The WSN about the healthcare domain can be broken into a three-tier system. As displayed from Fig. 7.1, we can observe that sensors form the basis of Tier 1. Any wearable sensor or ubiquitous sensor forms the basis of this domain. An electrocardiogram

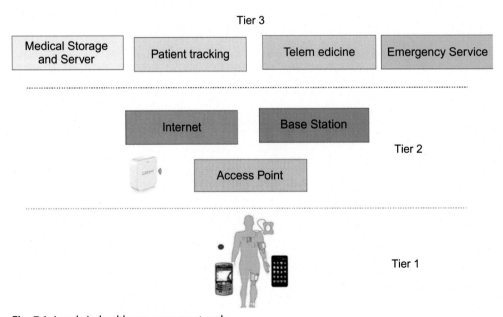

Fig. 7.1 Levels in healthcare sensor network.

(ECG) sensor, a wearable smartwatch having a photoplethysmography (PPG) sensor or pressure sensor, or any ambient sensor present in the environment, for example a temperature sensor can be part of this layer. Thus Tier 1 encompasses a wide number of wireless sensors that are merged to form the varying WSN. Using the proper communication technology, the collected raw data is relayed onto the next layer (Tier 2). The middle layer (Tier 2) is the backbone layer or the layer responsible for the actual delivery. The middle layer forms the wired or wireless network through which the data is passed onto the higher abstracted level. Tier 3 is the enterprise level where end users of the system can be a company, or for patient tracking and monitoring, emergency service, medical cloud server or for analysis, etc. The WSN framework [3] is built concerning the required application. For example, continuous monitoring in a hospital requires the usage of various biosensors that are deployed to collect information and relay accordingly. For companies and industries, the data are collected wirelessly from the smart band, smartwatches, smart shoes, etc. for various research and analysis purposes. In general, the data concerning health can be classified as periodic, aperiodic, sporadic, and one-shot data. Periodic signals repeat after a certain interval of time like ECG signal, respiratory signal, etc. Aperiodic signals [4] are signals that do not repeat after a certain interval of time. An ECG signal can also be an aperiodic signal if there is heart arrhythmia or heart problem. Sporadic signals are some random high or low values that are generated due to various events pertaining to health conditions or random noises involved during the data collection process. Understanding sporadic signals requires a very high level of analysis to differentiate between noise and actual data and what does the data even signify? In the case of one-shot data transmission, the collected data has to be sent immediately by all the deployed sensors in the network.

Wireless communication is the foundation on which the sensor network system is built. In WSN-based healthcare, only some of the appropriate standards are used. IEEE 802.11 is a standard for WLAN, providing continuous networking rather than coverage, mobility, or energy efficiency. WSN is a more resource-restricted network where the common IEEE 802.11 standard is not suitable for healthcare data transmission due to the various orientations in the design. Then we have the IEEE 802.20 standard which is built for mobile broadband services and is in direct competition to 3G technology of cellular platforms. IEEE 802.16 (WiMAX) is compatible with telemedicine-related service providers in both fixed and mobile environments. The WiMAX has numerous networking layers to provide greater coverage and internet bandwidth, very high-quality images like CT-SCAN, and USG images can be transmitted through this network with reduced delays. Thus the framework of communication supports diagnostics, analysis, and monitoring with the presence of a large network capability. There is also a very good quality of Services (QoS) that increases the efficiency and makes the system robust. This is much suitable for real-time services where the details about the patients can be sent to the doctor beforehand, for example from an ambulance. For Wireless Personal Area Network the most important standards covering link technologies between the Base Stations and coverage/relay nodes for healthcare application are based on IEEE 802.15. IEEE 802.15.1 is the Bluetooth technology which is also sometimes regarded as the long-term health-related access technology due to its high transmission power. Bluetooth low energy (BLE) is also very popular where the consumption of energy is very less but it gives range and services comparable to classical Bluetooth technology. BLE is implemented using two alternatives: (1) single mode and (2) dual mode. The power consumption in a single mode is very low, allowing them to run for a long time using a single battery source. The single-mode device has a very compact radio communication unit suitable for incorporating into the wireless medical monitors. The only problem with single-mode devices is that they cannot directly communicate with other regular Bluetooth devices. In dual mode, devices can communicate with dual-mode as well as single-mode devices and are majorly focused on radio communication targeted for PC or handheld devices. Medical data can be transferred from wireless sensors and can be relayed to a base station which can be a mobile device or a PC and further to a remote caregiver. In the majority of the WSN application Zigbee is used for its low cost and low power WSN. It is a very popular technology mostly used in acute hospitals. The only problem is the low data rate or bandwidth which is in kbps which is not suitable for sending large-size data over the network. In the healthcare WSN, it is important to focus on ubiquitous sensing due to the vast application domain; the system of healthcare is an amalgamation of various technologies to build the networking architecture about various application needs.

7.3 Transmission methods

The data collection and gathering mechanism are vividly discussed in Chapter 2 of the book. After the collection procedure, data are relayed toward a base station. The base stations are deployed in a manner to act as the supervising and collection node in a wireless sensor network (WSN). Base stations are deployed area-wise, and the sensors relay the information to its nearest base station. The communication protocol used in this scenario can be thought of as a three layer protocol stack where the network layer is in charge of routing the data forwarded by the transport layer. Data is traversed in a multi-hop wireless manner from one sensor node to the other sink. The data link layer is in charge of various operations of data link control. From multiplexing to frequency channel allocation to Media Access Control (MAC) as well as error control are performed by the data link layer. The sensor nodes are usually of two types:

(1) *Relay nodes*: Responsible for only relaying information from one node to the other.
(2) *Coverage nodes*: Responsible for both data gathering and relaying of data.

7.3.1 Unicast

The objective of the unicast transmission is to send data in a one-to-one delivery manner from one sensor node to another. One of the sensors is the sensor and the other one is a receiver. The sensors involved in such kind of transmission have a lot of limitation. The battery power of such sensors is low and has low processing capabilities. Usually, Time Division multiplexing is used as a media access control (MAC) protocol. The task of sensing has been divided into the network. The routing path is chosen by the network, the shortest path(s) is usually chosen for routing. Efficient unicast routing is given by Spachos et al. in [5] where the authors have proposed a cognitive unicast routing that discovers routes when where required. In the Unicast routing mechanism, the sensors or the relay nodes do not have the whole knowledge of the network. Thus an explicit route discovery is required when the transmission needs to take place.

7.3.2 Multicast

In multicast, transmission is required when a node needs to transmit data to specific nodes in the network. The data is meant only for the node to which the sender wants to send it. The in-between relay nodes only forward the information. During the multicast operation, it might be the case the same information gets routed in a different path where a loop is there, thus resulting in redundant information sharing to the same node in the network. Thus to eliminate the loop during the multicast operation, various path marker procedures are developed for routing that stores the history of packets consumed during the procedure. One such efficient multicast routing procedure is given by Chun et al. in [6].

7.3.3 Broadcast

In broadcast transmission [7], the nodes share the data with all the nodes in the network. In the traditional approach, the nodes broadcast the information by flooding the information to the adjacent nodes. The nodes that have received the data or packets broadcast the packets to the nodes that they are adjacent with. In this manner, all the nodes receive the data but traditional broadcasting suffers from redundant data reception. The sensor nodes are battery-operated devices so broadcast transmission must happen in an energy-efficient way. The internet in the present paradigm is an amalgamation of cellular networks, Wireless Sensor Network, Ad Hoc Network, and wired and wireless networks (Fig. 7.2).

7.3.4 Convergecast

Convergecasting is the procedure through which the data collected by the coverage nodes are relayed through the network toward a common sink node or base station. The data relaying can be applied as well as area-specific needs. For example, if one needs to collect the temperature data from a particular general ward of the hospital, a query can be submitted to a higher level application program, which will redirect the query and the

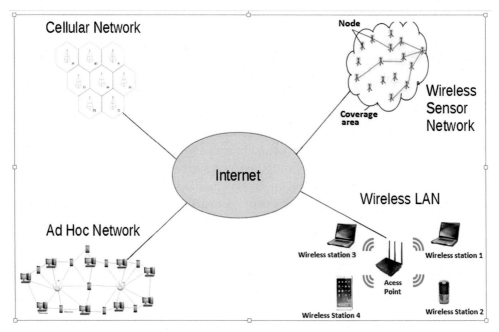

Fig. 7.2 Present internet overview.

sensors about the ward connected to the patients will be activated. The sensed data will be relayed back toward a common base station about the ward which will, in turn, forward the information back to the higher levels of the system. The term used for the collection of data is "aggregation." Data aggregation [8] is the concept in which the data collection is performed from all the sensors in an energy-efficient manner to increase the network lifetime. The sensors used in WSN are battery-operated devices, thus it is essential that the coverage nodes and relay nodes perform the respective operation concerning application-specific need. The data is gathered from various sensors in the network. Numerous procedures like low energy adaptive clustering hierarchy (LEACH) [9] and Tiny Aggregation (TAG) [10] are used to aggregate the sensors nodes' data. The input to the aggregation procedure is the data taken from various nodes and output is the aggregated data. It is also very important to select the path through which the data will be relayed toward the sink. Thus abstractly the aggregation [11] can be classified into following types: no aggregation, partial aggregation, and full aggregation.

In order to get an overview of the working of the aggregation process, consider Fig. 7.3. Nodes "*a*," "*k*," "*d*," "*e*," and "*f*" which are marked as red can be considered as coverage nodes or the nodes that gather data. While node "*c*" is a relay node that doesn't produce any extra data packets. The coverage nodes can be attached with any patient for monitoring temperature and other vitals. The gathered data is relay toward the base station "*B*" where the data are monitored, analyzed, or passed onto a cloud. Let us consider Time Division Multiplexing is the means of transmission and reception

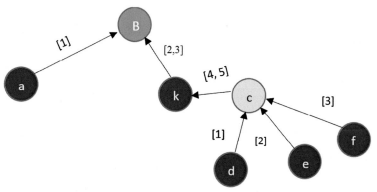

Fig. 7.3 Slot allocation in sensor network.

of packets. The TDMA protocol [12] is a channel access mechanism that allows several nodes to send and receive data by dividing the signals into different slots under the same frequency channel. Let us consider the length of the time slot to be two, that is every slot can carry two packets. As we can observe that node "a" sends the data directly to the base station "B" using the time slot 1, nodes "d," "e," and "f" are communicating using slot 1, 2, 3, respectively. Now, "c" being the relay node, relays the packet to node "k," but used two time slots 4, 5; it cannot use 1, 2, 3 as they are part of the adjacent data transmission from nodes "d," "e," and "f."

It might be the case the range of nodes "d" and "e" might overlap and they are sensing the same information which is relayed to node "c." Thus if no aggregation is considered then the number of packets, data redundancy, and packet retransmission will increase which will decrease the life of the network. Full aggregation, on the other hand, contributes to energy saving but suffers from delay during the transmission process. Concerning routing procedure, the aggregation method is classified as Tree-, Cluster-, and Network-based aggregation.

Tree-based approach

In tree-based aggregation [13], a tree is constructed from the base station to the underlying leaf sensor nodes. The aggregation tree (Fig. 7.4A) is a minimal spanning tree and the raw data is transferred from the source to the sink by passing through the relay nodes which are in turn the parent node of a node. The path is discovered with each relay operation from the node to its parent sensor node. Usually, in tree-based aggregation, the leaf nodes are coverage nodes and thus one has to ensure that no relay nodes are part of the leaf node-set. The aggregation is done by the parent nodes on the reception of data packets from the underlying leaf nodes at each level.

Cluster-based aggregation

In a cluster-based approach [14] (Fig. 7.4B), the data aggregation is performed on multiple regions of interest. Thus each location is divided as a separated cluster which can employ a tree-based approach underneath or another approach for the

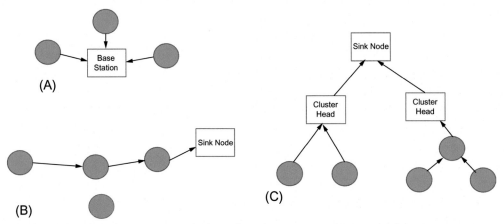

Fig. 7.4 (A) Tree-based aggregation. (B) Cluster-based aggregation. (C) Network-based aggregation.

aggregation procedure. Each cluster has a cluster head or the master node that aggregates the information received from the underlying nodes of the respective cluster. The aggregated information is further relayed to the main base station. Energy consumption is reduced as the network is clustered and the required data relaying doesn't take place directly to the base station but happens via the cluster head.

Network–based aggregation

Network-based aggregation [11] (Fig. 7.4C) aggregates by reducing the size of the packets and thereby reducing the energy consumption of the network. The nodes in between or the intermediate nodes are responsible for the aggregation process. The aggregation is done by performing some aggregate function over the gathered data. A simple overview of aggregation function over periodic data can be as follows. For example, for patient monitoring, periodic temperature data is gathered from multiple patients. The intermediate nodes aggregate the data by sending only the temperature if it crosses a certain threshold of 99°C or if the temperature is greater than the previously gathered temperature. Thus the number of packets sent gets decreased hence the life of the network increases.

7.4 Cloud computing

The growth of data produced by various sensors in the healthcare domain and medical community requires efficient storage, analysis, sharing, and resource management. Cloud computing [15] is one of the important and most plausible solutions which supports the healthcare infrastructure and helps in support new infrastructure. Cloud computing enables ubiquitous and on-demand processing to a shared pool of resources that can be quickly equipped and deployed very easily. The basic cloud computing architecture consists of four layers. At the bottom-most layer lies the actual server where the data are

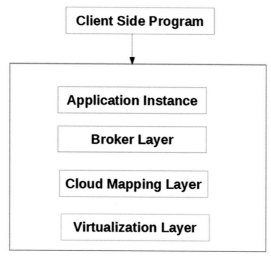

Fig. 7.5 Basic components of cloud computing.

kept for storage. The system performs in a manner that the whole processing from storage to accessing and analysis is abstracted from the user. The user doesn't know which system the respective data is kept in. Further, the whole system is distributed hence the very same data is replicated across multiple servers to ensure robustness and sharing. At the client's end, the users use the system via the client application. The four basic parts that constitute a cloud are shown in Fig. 7.5. The user's application instances are created in the instance layer. The cloud broker is an intermediate layer between the users and the service providers. A cloud broker [16] is responsible for various operations like server load distribution, partnering with more than one cloud service provider, etc. The virtualization layer is responsible for creating a virtual version of resources and services. Virtualization involves creating a specialized software version of the resources upon which various operating can be executed to manage the work on the same infrastructure. The three popular types of cloud are as follows.

Private cloud

The private cloud infrastructure [17] is not openly accessible to everyone. The system is distributed in nature like any cloud architecture and is hosted privately in the organization's data center or externally which is called an outsourced private cloud. Special emphasis on security is given to protect the integrity of data about the enterprise. It is mainly suitable for enterprises that require full control over the cloud infrastructure pertaining to their organization's needs.

Public cloud

The public cloud [18] is open to everyone on the internet for accessing, storing, and using cloud computing power for any computation provided by the cloud service provider. Some of the notable service providers are Amazon EC2, Microsoft, and Google providing easier access, integration, and flexibility to all the users. The public cloud is highly scalable and there is no limit to the number of users connected to the system.

Hybrid cloud

Hybrid cloud [19] brings the integration between the public as well as the private cloud infrastructure with the objective of sharing the computational load and data among both the infrastructures. In a hybrid cloud infrastructure services provided by the public cloud is available for access by all, while the private cloud infrastructure is still private to the organization. The hybrid brings in infrastructure for the organization which requires some part of its data to be kept in a more secure facility.

Community cloud

In community cloud [20], systems are owned by more than one organization and the services are shared and accessible only among the community organization. For example, the cloud infrastructure pertaining to the research community or medical community where some of the parts of the data are shared publicly while other sensitive information is kept in private infrastructure.

In base level of any computing paradigm, it consists of a collection of interconnected distributed computers which are virtualized in the higher level for dynamic provisioning. Usually, the cloud computing domain is categorized into three types of models:

(i) *Infrastructure as a service*: In infrastructure as a service (IaaS) [21] the resources and computing power are provisioned on demand. In IaaS the user doesn't need to maintain the physical server, and the costing is done on a pay-as-you-go basis.

(ii) *Platform as a service*: Platform as a Service (PaaS) [22] provides resources such as operating systems and various frameworks that allow individual users to create their applications and deploy them. It is accessible to users easily and the web services and databases are integrated with the cloud platform. Some of the examples of *PaaS* are Google App Engine, Windows Azure, etc.

(iii) *Software as a service*: The objective of software as a service (SaaS) [23] is to provide on-demand software applications that provide cloud computing power to the users with the objective of storage, sharing as the sole focus of the system. Some of the popular SaaS are Dropbox, Google Drive, ZenDesk, etc.

7.5 Medical cloud application

7.5.1 Information management and sharing

Integrating cloud computing into the healthcare infrastructure allows the management and sharing of healthcare resources. Cloud computing platforms provide unified patient data collected across numerous operators. As the data is stored over a distributed platform, the sharing of information becomes easier. The size of medical image data is quite large, various image archives are provided by cloud service providers. The Digital Imaging and Communications in Medicine (DICOM) server is included in the cloud computing platform that handles standard operations such as searching, accessing and retrieval, and storing queries. DICOM [24] is a standard for the retrieval and storage of image data. DICOM has an indexing program that parses the metadata and stores them into an SQL server. Based on various patient metadata and image attributes the images are searched across the database. Cloud is also used for storing and managing clinical data gathered via various sensors attached to the medical monitoring equipment. The gathered information is analyzed by medical practitioners for further processing. Mohapatra et al. in [25] proposed a cloud-based framework for ubiquitous data gathering and storing of clinical information in the cloud with real-time monitoring capabilities.

7.5.2 Medical support system

The integration of the cloud with the monitoring of patients in critical care is built on the foundation of the wireless sensor network. The gathered sensor data from various patients are routed and stored into an emergency monitor system which is present in the cloud. Mohapatra et al. in [26] give a telemedicine-oriented study. Telemedicine is the delivery of medical services like tests, consultation, etc. over to patients remotely. The work presents a framework for sensor cloud infrastructure where the query received from the patient monitoring system is relayed through the sensor network in the cloud interface. The cloud is responsible for executing the query and sharing the information among medical practitioners, caretakers, and pharmacies. During emergency management, past patient-centric data is important which helps in better diagnosis (Fig. 7.6A) of the disease and brings the whole medical system into streamline, minimizing any error in the treatment. In [27] an Emergency Management System (EMS) that integrates cloud

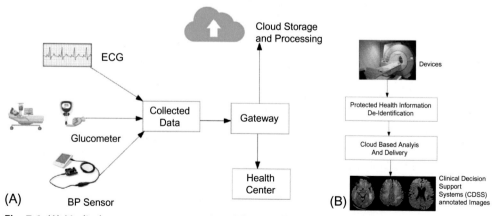

Fig. 7.6 (A) Medical support system overview. (B) Image-based clinical analysis.

infrastructure for information sharing has been proposed. The work identifies emergency departments such as emergency ambulance communication, Ambulance service, and Ambulance physician, Hospital, etc., involved during emergency medical scenarios. To improve the quality of care, the information about each of the departments is processed and shared with the other departments for effective processing.

7.5.3 Clinical analytics

In the present era, statistics and machine learning-based analysis are widely used in numerous disease diagnoses and user-centric analyses. Powerful Deep Learning models have been trained for image-based analysis. Instead of developing a dedicated infrastructure, cloud-based analysis is widely chosen where the base station at the clinical end collects all the data and submits the data as input to the cloud through a proper communication gateway. A cloud-based framework for image analytics is proposed in [28,29] where the DICOM medical images (Fig. 7.6B) collected for medical purposes are shared with radiologists and organizations for image analysis and disease diagnosis remotely. Before the DICOM images are shared with third parties, strict guidelines are followed concerning the de-identification of Protected Health Information (PHI) from the medical data.

7.6 Issues in healthcare cloud

The integration of cloud computing with healthcare has paved the way for various applications in the medical domain. Although cloud computing comes as a blessing it has its share of issues.

(1) **Interoperability**: The healthcare cloud contains the integration of various organizations involving various services from multiple cloud service providers. In order to achieve proper integration between client ends, clouds, and services, the use of a fixed framework and APIs is very important. The need for good interoperability is important in order to achieve easy data migration.

(2) **Cloud maintenance**: As the size of the cloud scales with the integration of multiple applications and user needs, the task of maintainability of the cloud platform becomes difficult.

(3) **Organizational issues**: Every organization is built on the very foundation of its policies and culture. Integrating cloud computing paves the path toward adopting newer technologies, thus how the employees will be affected by the change should also be considered. A easier transition should be adopted in the healthcare organizations. Thus the adoption of newer workflows and policies should be considered.

(4) **Constraint in regulation**: The integration of cloud computing with the healthcare domain should comply with the various policies. For example, the Personal

Information Protection and Electronic Documents Act (PIPEDA) and Health Insurance Portability and Accountability Act (HIPAA) state that the data about healthcare should comply with some prescribed means. Furthermore, the rules and regulations about the various countries where the cloud is integrated should be followed.

7.7 Conclusion

The chapter gives insights into the wireless sensor network domain and how it can be utilized in clinical analytics. The chapter also discusses the data aggregation methodologies used in the wireless sensor network. It can be observed that the integration of cloud computing into the healthcare domain has opened up a plethora of opportunities paving toward easy information sharing to complicated analysis.

References

[1] M.E.S.U.T. Çiçek, Wearable technologies and its future applications, Int. J. Electr. Electron. Data Commun. 3 (4) (2015) 45–50.

[2] J. Yick, B. Mukherjee, D. Ghosal, Wireless sensor network survey, Comput. Netw. 52 (12) (2008) 2292–2330.

[3] K. Haseeb, I. Ud Din, A. Almogren, N. Islam, An energy efficient and secure IoT-based WSN framework: an application to smart agriculture, Sensors 20 (7) (2020) 2081.

[4] S. Zhao, Y. Peng, F. Yang, E. Ugur, B. Akin, H. Wang, Health state estimation and remaining useful life prediction of power devices subject to noisy and aperiodic condition monitoring, IEEE Trans. Instrum. Meas. 70 (2021) 1–16.

[5] P. Spachos, P. Chatzimisios, D. Hatzinakos, Energy efficient cognitive unicast routing for wireless sensor networks, in: 2013 IEEE 77th Vehicular Technology Conference (VTC Spring), IEEE, 2013, June, pp. 1–5.

[6] Z. Yi, L. Yan-Chun, An improved ant colony optimisation and its application on multicast routing problem, Int. J. Wirel. Mob. Comput. 5 (1) (2011) 18–23.

[7] D. Yubo, N. Jianwei, L. Lian, Modeling of broadcasting based on distance scheme for WSN, in: 2009 Fifth International Joint Conference on INC, IMS and IDC, IEEE, 2009, August, pp. 1176–1179.

[8] R. Rajagopalan, P.K. Varshney, Data Aggregation Techniques in Sensor Networks: A Survey, 2006.

[9] S.K. Singh, P. Kumar, J.P. Singh, A survey on successors of LEACH protocol, IEEE Access 5 (2017) 4298–4328.

[10] S. Madden, M.J. Franklin, J.M. Hellerstein, W. Hong, TAG: A tiny aggregation service for ad-hoc sensor networks, ACM SIGOPS Oper. Syst. Rev. 36 (SI) (2002) 131–146.

[11] G. Dhand, S.S. Tyagi, Data aggregation techniques in WSN: survey, Procedia Comput. Sci. 92 (2016) 378–384.

[12] V. Cionca, T. Newe, V. Dadârlat, TDMA protocol requirements for wireless sensor networks, in: 2008 Second International Conference on Sensor Technologies and Applications (sensorcomm 2008), IEEE, 2008, August, pp. 30–35.

[13] O.D. Incel, B. Krishnamachari, Enhancing the data collection rate of tree-based aggregation in wireless sensor networks, in: 2008 5th Annual IEEE Communications Society Conference on Sensor, Mesh and Ad Hoc Communications and Networks, IEEE, 2008, June, pp. 569–577.

[14] V.S. Devi, T. Ravi, S.B. Priya, Cluster based data aggregation scheme for latency and packet loss reduction in WSN, Comput. Commun. 149 (2020) 36–43.

[15] K. Ahmed, M. Gregory, Integrating wireless sensor networks with cloud computing, in: 2011 Seventh International Conference on Mobile Ad-hoc and Sensor Networks, IEEE, 2011, December, pp. 364–366.

[16] S.G. Grivas, T.U. Kumar, H. Wache, Cloud broker: bringing intelligence into the cloud, in: 2010 IEEE 3rd International Conference on Cloud Computing, IEEE, 2010, July, pp. 544–545.

[17] F. Doelitzscher, A. Sulistio, C. Reich, H. Kuijs, D. Wolf, Private cloud for collaboration and e-Learning services: from IaaS to SaaS, Computing 91 (1) (2011) 23–42.

[18] A. Li, X. Yang, S. Kandula, M. Zhang, CloudCmp: comparing public cloud providers, in: Proceedings of the 10th ACM SIGCOMM Conference on Internet Measurement, 2010, November, pp. 1–14.

[19] Q. Li, Z.Y. Wang, W.H. Li, J. Li, C. Wang, R.Y. Du, Applications integration in a hybrid cloud computing environment: Modelling and platform, Enterp. Inf. Syst. 7 (3) (2013) 237–271.

[20] A. Marinos, G. Briscoe, Community cloud computing, in: IEEE International Conference on Cloud Computing, Springer, Berlin, Heidelberg, 2009, December, pp. 472–484.

[21] A. Iosup, R. Prodan, D. Epema, Iaas cloud benchmarking: approaches, challenges, and experience, in: Cloud Computing for Data-Intensive Applications, Springer, New York, NY, 2014, pp. 83–104.

[22] C. Pahl, Containerization and the PaaS cloud, IEEE Cloud Comput. 2 (3) (2015) 24–31.

[23] M. Cusumano, Cloud computing and SaaS as new computing platforms, Commun. ACM 53 (4) (2010) 27–29.

[24] P. Mildenberger, M. Eichelberg, E. Martin, Introduction to the DICOM standard, Eur. Radiol. 12 (4) (2002) 920–927.

[25] S. Mohapatra, K. Smruti Rekha, Sensor-Cloud: A Hybrid framework for remote patient monitoring, Int. J. Comput. Appl. (0975–8887) 55 (2) (2012).

[26] V. Koufi, F. Malamateniou, G. Vassilacopoulos, Ubiquitous access to cloud emergency medical services, in: Information Technology and Applications in Biomedicine (ITAB), 2010 10th IEEE International Conference on, pp. 1–4, 3–5 Nov, 2010.

[27] L. Ehwerhemuepha, G. Gasperino, N. Bischoff, S. Taraman, A. Chang, W. Feaster, HealtheDataLab–a cloud computing solution for data science and advanced analytics in healthcare with application to predicting multi-center pediatric readmissions, BMC Med. Inf. Decis. Mak. 20 (1) (2020) 1–12.

[28] I.H. Arka, K. Chellappan, Collaborative compressed I-cloud medical image storage with decompress viewer, Procedia Comput. Sci. 42 (2014) 114–121.

[29] C.C. Teng, J. Mitchell, C. Walker, A. Swan, C. Davila, D. Howard, T. Needham, A medical image archive solution in the cloud, in: 2010 IEEE International Conference on Software Engineering and Service Sciences, IEEE, 2010, July, pp. 431–434.

Index

Note: Page numbers followed by *f* indicate figures, *t* indicate tables and *b* indicate boxes.

Printed in the United States
by Baker & Taylor Publisher Services